D1067358

Flexagons Inside Out

LESLIE PHILIP (Les) POOK was born in Middlesex, England in 1935. He obtained a BSc in metallurgy from the University of London in 1956. He started his career at Hawker Siddeley Aviation Ltd, Coventry, in 1956. In 1963 he moved to the National Engineering Laboratory, East Kilbride, Glasgow. In 1969, while at the National Engineering Laboratory he obtained a PhD in mechanical engineering from the University of Strathclyde. Dr Pook moved to University College London in 1990. He retired formally in 1998 but remained affiliated to University College London as a Visiting Professor. He is a Fellow of the Institution of Mechanical Engineers and a Fellow of the Institute of Materials, Minerals and Mining.

Dr Pook has wide experience of both research and practical engineering problems involving metal fatigue and fracture mechanics. In these fields he has published four books and over a hundred papers. Present professional interests include the fatigue behaviour of cracks in complex stress fields, finite element analysis, and language editing of translated technical material. His leisure activities have for many years included recreational mathematics, horology, in which he has published one paper, and gardening. He regards his extensive DIY activities, especially plumbing, as a chore.

Les married his wife, Ann, in 1960. They have a daughter, Stephanie, and a son, Adrian.

Flexagons
Inside Out

Les Pook

University College London

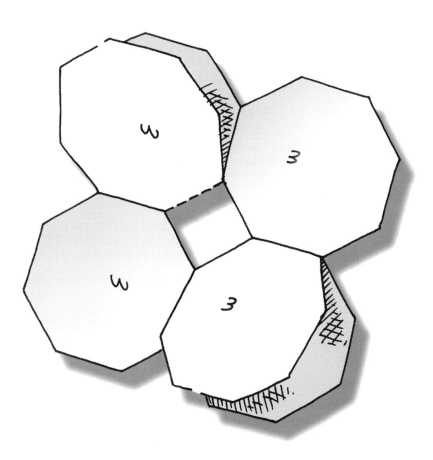

CAMBRIDGE
UNIVERSITY PRESS

PUBLISHED BY THE PRESS SYNDICATE OF THE UNIVERSITY OF CAMBRIDGE
The Pitt Building, Trumpington Street, Cambridge, United Kingdom

CAMBRIDGE UNIVERSITY PRESS
The Edinburgh Building, Cambridge CB2 2RU, UK
40 West 20th Street, New York, NY 10011-4211, USA
477 Williamstown Road, Port Melbourne, VIC 3207, Australia
Ruiz de Alarcón 13, 28014 Madrid, Spain
Dock House, The Waterfront, Cape Town 8001, South Africa

http://www.cambridge.org

© Cambridge University Press 2003

First published 2003

Printed in the United Kingdom at the University Press, Cambridge

Typefaces Swift 10/14 pt   *System* LaTeX 2_ε [TB]

A catalogue record for this book is available from the British Library

ISBN 0 521 81970 9 hardback
ISBN 0 521 52574 8 paperback

Contents

Preface

Mathematics, rightly viewed, possesses not only truth, but supreme beauty –
a beauty cold and austere, like that of sculpture.

Bertrand Russell, *Mysticism and Logic.*

Flexagons are hinged polygons that have the intriguing property of displaying different pairs of faces when they are 'flexed'. Workable paper models of flexagons are easy to make, and entertaining to manipulate. Flexagons have a surprisingly complex mathematical structure and just how a flexagon works is not obvious on casual examination of a paper model. The aesthetic appeal of flexagons is in their dynamic behaviour rather than the static appeal of, say, polyhedra. One of the attractions of flexagons is that it is possible to explore their dynamic behaviour experimentally as well as theoretically. Flexagons may be appreciated at three different levels: firstly as toys or puzzles, secondly as a recreational mathematics topic, and finally as the subject of serious mathematical study. Well made models of simple flexagons can be bought, but they are not widely available.

I first became interested in flexagons in the late 1960s through reading two of Martin Gardner's books. At that time I made numerous paper models of flexagons, and carried out some theoretical analysis. I did try writing up some of the material for publication, but editors showed a marked lack of enthusiasm and I lost interest. In 1997, while browsing in a library I came across a paper on flexagons, and this revived my interest. Fortunately, my collection of flexagon models and associated notes had survived three house moves.

I carried out a literature search as part of my renewed investigations. This showed that there is only a limited amount of published information on flexagons. Furthermore, this information is scattered and some items are difficult to locate. There may well be significant items which I have missed. The only books I found devoted entirely to flexagons are at the toy or puzzle level. These usually contain cut out nets for paper models of flexagons, but making these up destroys the book. At the recreational mathematics level I found several books containing sections or chapters on flexagons, but the information included tends to be fragmentary, and is sometimes difficult to appreciate fully without some prior knowledge of flexagons. At this level I did not find any books devoted entirely to flexagons.

The first paper at the serious mathematics level was published in 1957 and a comprehensive report on flexagons was issued in 1962. I only recently obtained a copy of this report and it turned out to include quite a lot of material which I had rediscovered for myself. At this level the subsequent literature is sparse. The main features of flexagons have been understood for half a century, but a long standing, and still not fully resolved, problem is how best to describe their structure and dynamic behaviour. For this reason publications on flexagons at the serious mathematics level tend either to be not very informative or to need close reading by a competent mathematician in order to fully appreciate their content.

Contemplation of my collection of publications on flexagons, together with the results of my own early and recent investigations, showed that I had enough information to write a book on flexagons at the recreational mathematics level. Helpful comments made by anonymous reviewers of the first draft of the book encouraged me to proceed. This book is the result. It is assumed that the reader has some knowledge of elementary geometry. No previous knowledge of flexagons is assumed, so the book is suitable as an introduction to flexagons at the toy or puzzle level. In general, detailed proofs are long and tedious so they are not included. Where there is uncertainty over the accuracy of a conclusion this is made clear in the text.

There is an infinite number of possible types of flexagon so no book on flexagons can be comprehensive. The material included is very much a personal choice. It is arranged roughly in order of increasing difficulty rather than in a strictly logical order which would be appropriate for a formal textbook. The terminology used is mostly based on that used by previous authors but to keep the text concise it was found that some new terms were needed. Specialised terms are listed in the index so that definitions can be easily located.

A feature of the book is a collection of nets, with assembly instructions, for a wide range of paper models of flexagons. They are printed full size and laid out so that they can be photocopied. Some of the flexagons are difficult either to make or to manipulate, and this is noted in captions. The flexagons have been chosen to complement the text. Some of the nets were specially designed using methods described in the book.

Notation

c Number of sides on the inscribed polygon of a flexagon figure.

f Number of faces on a flexagon.

n Number of sectors on a main position of a flexagon.

$2n$ Number of polygons on a face of a main position of a flexagon. Number of polyhedra on a hyperface of a main position of a flexahedron.

s Number of sides on a polygon. Number of sides on the circumscribing polygon of a flexagon figure.

$\{s\}$ Schläfli symbol for a regular polygon.

$\langle s, c \rangle$ Flexagon symbol for a regular flexagon.

Introduction

Chapter 1 is an introductory chapter. Nets and assembly instructions are given for a simple hexaflexagon, the trihexaflexagon, and for a simple square flexagon. The pinch flex used to manipulate them is described. Nets for other types of flexagon are given later in the book to illustrate various points made. General assembly instructions are given for these nets.

Flexagons are a twentieth century discovery. Their early history is given in Chapter 2. In 1940 two members of a Flexagon Committee at Princeton University worked out a mathematical theory of flexagons but this was never published. The subject can be said to have reached maturity with the issue in 1962 of a comprehensive report on flexagons, but it was not published in a form which reached a wide audience.

In general the main characteristic feature of a flexagon is that it has the appearance of a polygon which may be flexed in order to display pairs of faces, around a cycle, in cyclic order. Another characteristic feature is that faces of individual polygons, known as leaves, which make up a face of a flexagon, rotate in the sense that different vertices move to the centre of a main position as a flexagon is flexed from one main position to another. The visible leaves are actually folded piles of leaves, called pats. Sometimes pats are single leaves. Alternate pats have the same structure. A pair of adjacent pats is a sector. A convenient mathematical framework for the analysis of flexagons is presented in Chapter 3, together with explanations of special technical terms. A straightforward geometric approach, without equations, is used. Geometric descriptions are used for three main purposes: firstly, to map the dynamic behaviour of flexagons, secondly to analyse their structure, and thirdly as the basis of recipes for the construction of flexagons of any desired type.

Hexaflexagons, described in Chapter 4, were the first variety of flexagon to be discovered and they have been analysed in the most detail. The leaves of a hexaflexagon are equilateral triangles. In appearance a main position of a hexaflexagon is flat and consists of six leaves, each with a vertex at the centre so there are six pats and three sectors. The outline is a regular hexagon. In some ways hexaflexagons are the simplest type of flexagon. There are only one possible type of cycle and one possible type of link between cycles. Multicycle hexaflexagons have been extensively analysed and their dynamic behaviour is well understood. There has been much interest in the design of nets for specific types of hexaflexagon whose dynamic behaviour is known. Design methods are described in detail. Extensive analysis has resulted in variations on the theme of hexaflexagons. Three are described in Chapter 5. These are a different variety of flexagon, triangle flexagons, a different way of flexing hexaflexagons, the V-flex, and origami like recreations with hexaflexagons.

Square flexagons are described in Chapter 6. They were the second variety of flexagon to be discovered. They are less well understood than hexaflexagons, partly because their dynamic behaviour is much more complex. Square flexagons have three different types of cycle and two types of link between cycles are possible. The leaves of a square flexagon are squares. In appearance a main position of a square flexagon is flat and consists of four leaves each with a vertex at the centre so there are four pats and two sectors. The outline is a square. The design of nets for specific types of square flexagons whose dynamic behaviour is known is more difficult than for hexaflexagons.

The remaining chapters are more advanced. Chapter 7 is an introduction to convex polygon flexagons. Convex polygon flexagons are generalisations of the square flexagons and triangle flexagons described in earlier chapters. Understanding of convex polygon flexagons in general is incomplete. There is an infinite family of convex polygon flexagons. Varieties are named after the constituent polygons. A feature of some varieties of convex polygon flexagon is that there may be more than one type of main position and more than one type of complete cycle. It then becomes necessary to refer to principal and subsidiary main positions and cycles. In a principal main position a convex polygon flexagon has the appearance of four leaves each with a vertex at the centre so there are four pats and two sectors. If a flexagon is regarded as a linkage then bending the leaves during flexing is not permissible. However, allowing bending during flexing does makes it easier to rationalise dynamic behaviours of the convex polygon flexagon family, and does make the manipulation of some types of convex polygon flexagon more interesting.

The first variety of the convex polygon flexagon family, the digon flexagon, can only be flexed in truncated form and then only by bending the leaves of a paper model using a push through flex. The second and third varieties are triangle flexagons and square flexagons. None of these first three varieties is typical of convex polygon flexagons. The fourth variety, pentagon flexagons, and higher varieties have characteristics in common and all can be regarded as typical members of the family. In a typical convex polygon flexagon the sum of the leaf vertex angles at the centre of a principal main position is greater than $360°$ so the principal main position is skew and its outline is a skew polygon. It is always possible to traverse the principal cycle of a typical convex polygon flexagon without bending the leaves of a paper model. There is always at least one subsidiary cycle. In general subsidiary cycles cannot be traversed without bending leaves. The appearance of the subsidiary main positions is different from that of the principal main positions. Various features of typical convex polygon flexagons are illustrated in Chapter 8 through descriptions of pentagon flexagons, hexagon flexagons and octagon flexagons. With octagon flexagons an additional type of flex, the twist flex, appears.

In a systematic treatment flexagons can be classified into two main infinite families. The first is the convex polygon flexagon family and the second is the star flexagon family. A principal main position of a star flexagon is flat, and has the appearance of an even number of regular polygons arranged about its centre, each with a vertex at the centre. The first two varieties of star flexagons are square flexagons and hexaflexagons. These are not typical of star flexagons. Typical star flexagons have at least eight polygons arranged about the centre of a principal main position, and the constituent polygons are regular star polygons. Interpenetration of the stellations during flexing makes the construction of paper models impossible.

Typical star flexagons are precursors to ring flexagons, which are described in Chapter 9. If all the stellations are removed from the constituent polygons of a star polygon flexagon then it becomes a ring flexagon. A principal main position of a ring flexagon has the appearance of a flat ring of an even number of regular convex polygons. The rings are regular in that each polygon is the same distance from the centre of the ring. Paper models of ring flexagons are awkward and tedious to handle. A compound flexagon is a ring flexagon in which alternate pats lie closer to the centre of a main position than do the others. There is an infinite number of compound flexagon varieties. Principal main positions are flat and have the appearance of compound rings of regular convex polygons in which alternate polygons lie closer to the centre of the ring. The leaves are regular convex polygons, and compound flexagons are named after the constituent polygons. The lines of hinges between pats do not intersect at the centres of the rings. Because of this, compound flexagons can only be flexed by bending the leaves. Flexing paper models is difficult.

Some distorted polygon flexagons are described in Chapter 10 in order to illustrate the enormous range of possibilities. A distorted polygon is a convex polygon derived from a regular convex polygon by changing the shape without changing the number of sides. The leaves of flexagons can be made from any convex polygon, but only a limited range of distorted convex polygons result in flexagons whose paper models are reasonably easy to handle. Distorted polygon flexagons are usually named after the polygons from which they are made. There are several ways in which the leaves of a flexagon made from regular polygons can be modified to produce a distorted polygon flexagon. A distorted polygon can sometimes be regarded as a partially stellated version of a regular convex polygon with a different number of sides. Alternatively, a distorted polygon can sometimes be regarded as a star polygon from which some of the stellations have been removed. If the proportions of the leaves are changed without changing the angular relationships between their sides then, in general, the dynamic behaviour of the flexagon is not affected. Changing the angular relationships between leaf sides does change the dynamic behaviour. Most distorted polygon flexagons are best regarded as variants of either convex polygon flexagons or star flexagons.

Four dimensional space is a purely theoretical idea but is nevertheless fascinating. Chapter 11 is a brief introduction to the remarkably rich and largely unexplored topic of flexahedra, which are the four dimensional analogues of flexagons. It is of course not possible to make physical models of flexahedra. It is possible to generate a flexahedron analogue of any flexagon and examples are given. There are some flexahedra which are not analogues of flexagons, and one is described. The nets of flexahedra are three dimensional so can be visualised in ordinary space. Sometimes main and intermediate positions of flexahedra, including those of the examples, are also three dimensional and hence may be visualised.

1 Making and flexing flexagons

As an introduction, nets and assembly instructions are given for two simple flexagons. The nets are laid out full size in a form suitable for photocopying. Nets for other types of flexagon are given later in the book to illustrate various points made. General assembly instructions are given for these nets. The appearance of paper models of flexagons can be improved by colouring and decorating the faces. Some decorative schemes exploit symmetries of flexagons both to create an attractive appearance and to create puzzles.

The 'pinch flex' used to manipulate flexagons in order to display different pairs of faces is described. A flexagon is flexed from a main position first to an 'intermediate position', and then to another main position. Other types of flex are sometimes used and are described later in the book.

1.1 The trihexaflexagon

The net for a paper model of a simple hexaflexagon is shown in Fig. 1.1. This is the trihexaflexagon, which was the first type of flexagon to be discovered. The trihexaflexagon is the simplest possible type of hexaflexagon (Conrad 1960, Conrad and Hartline 1962, Cundy and Rollett 1981, Gardner 1965, Gardner 1988, Hilton and Pedersen 1994, Hilton et al. 1997, Johnson 1974, Kenneway 1987, Laithwaite 1980, Liebeck 1964, McIntosh 2000h, McLean 1979, Madachy 1968, Maunsell 1954, Mitchell 1999, Oakley and Wisner 1957, Pedersen and Pedersen 1973, Wheeler 1958).

To make the trihexaflexagon photocopy the net onto 80 g/m^2 paper and cut it out. Crease the lines between triangles to form hinges. Transfer each number in brackets to the reverse face of the triangle, and delete it from the upper face. Fold the faces of triangles numbered 3 together. Join the ends of the net, shown by dashed lines on the figure, with transparent adhesive tape. The assembled trihexaflexagon is a continuous band of hinged triangles. The outline of the assembled trihexaflexagon is a hexagon, and it is in a 'main position'. One of the visible faces consists of six triangles numbered 1, each with a vertex at the centre of the hexagon. On the other visible face the triangles are numbered 2. This particular flexagon is a twisted band, so it exists in two mirror image (enantiomorphic) forms. To make the enantiomorph transfer each unbracketed

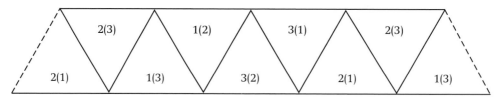

Fig. 1.1   Net for the trihexaflexagon.

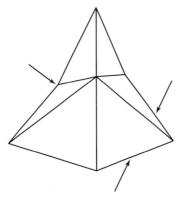

Fig. 1.2   Flexing a hexaflexagon using the pinch flex. Threefold rotational symmetry is maintained during flexing.

number to the reverse face of the triangle, and delete it from the upper face.

The individual polygons used to make flexagons, in this case triangles, are called 'leaves'. Examination of the assembled trihexaflexagon shows that it consists of alternate single leaves and folded piles of two leaves. Both the single leaves and the folded piles are called 'pats'.

1.2   The pinch flex

To 'flex' the trihexaflexagon start with the face numbered 1 uppermost. Pinch together two pats with the thumb and index finger of one hand, as shown in Fig. 1.2. Ensure that there isn't a continuous fold connecting the top of the two pats being pinched together. If there is, rotate the trihexaflexagon through 60° and start again. At the same time push the two opposite pats inwards with the index finger of the other hand. Continue until the trihexaflexagon has the appearance of three triangles connected at a common edge, and it is in an 'intermediate position'. The intermediate position has threefold rotational symmetry with an angle of 120° between each pair of triangles. For a description of various types of symmetry see Holden (1991). At the intermediate position it is possible to open the trihexaflexagon at the top of the common edge to reveal leaves numbered 3. Do this, and then flatten the trihexaflexagon. It will then have leaves numbered 3 on top and those numbered 1 underneath, and is in another main position. This manoeuvre is a 'pinch flex'. A pinch flex has two stages. In the first stage a main position is transformed into an intermediate position, and in the second stage this intermediate is transformed into another main position. Threefold rotational symmetry is maintained while pinch flexing the trihexaflexagon.

Repeating the pinch flex, after rotating the trihexaflexagon through 60°, results in leaves numbered 2 on top and those numbered 3 underneath. A third pinch flex returns the trihexaflexagon to its initial main position with 1 on top and 2 underneath, so completing a cycle. This cycle can be repeated indefinitely. The effect is that the band of triangles is continually turned inside out. The cycle can be traversed in the reverse direction by turning the trihexaflexagon over.

The pinch flex is the basic flex used to manipulate flexagons. It differs in detail for flexagons made from other types of polygon. Other types of flex are described later.

1.3   A simple square flexagon

In a main position a square flexagon has the appearance of four squares, each with a vertex at the centre. The outline is a larger square. Fig. 1.3 shows the net for a simple square flexagon (Chapman 1961, Conrad 1960, Conrad and Hartline 1962, Gardner 1966, Johnson 1974, McIntosh 2000c, McIntosh 2000g, Mitchell 1999, Neale 1999).

This square flexagon is assembled as for the trihexaflexagon (Section 1.1), except that only the horizontal lines between leaves need to be creased. The assembled square flexagon is a twisted band of hinged squares and it exists in two enantiomorphic forms. It is flexed using the pinch flex. This is simpler than for the trihexaflexagon because there is only twofold rotational symmetry. Start with the leaves numbered 2 uppermost. Fold the square flexagon in two with the fold uppermost. Ensure that there isn't a continuous fold connecting the tops of the pats being folded together. The square flexagon is now in an intermediate position. This has the appearance of two squares with a common side. Open the square flexagon at the top to reveal the squares numbered 3 and flatten it to a second main position. This particular square flexagon cannot be made to traverse a cycle. It can be returned to its initial main position by turning it over and flexing it. The version of the pinch flex used to manipulate square flexagons is sometimes called the 'book flex'.

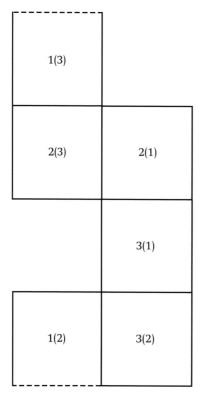

Fig. 1.3   Net for a simple square flexagon.

1.4   General assembly instructions

To assemble nets given later for other types of flexagons use the following general scheme. Copy a net onto 80 g/m^2 paper and cut it out. Crease the lines between leaves to form hinges. Ensure that adjacent leaves superimpose neatly when folded together. Transfer each number in brackets to the reverse face of the leaf, and delete it from the upper face. Copy any hinge or vertex letters on to the reverse of the net. Leaves with the same numbers are folded together. Start with the highest number and work downwards until only leaves numbered 1 and 2 are visible. Then join the ends of the net using transparent adhesive tape. Alternative or additional instructions are included in the captions for some nets.

Sometimes paper models of flexagons don't flex smoothly. If this is a problem try trimming a small amount, say 1 mm, off the edges of the net. Where a paper model of a flexagon is inherently difficult to make or to flex this is noted in the caption. Most types of flexagon are twisted bands and hence exist as an enantiomorphic pair. The net for one enantiomorph can be converted into the net for the other by reversing the

Fig. 1.4   Paper for flexagons made from equilateral triangles.

Fig. 1.5   Paper for flexagons made from squares.

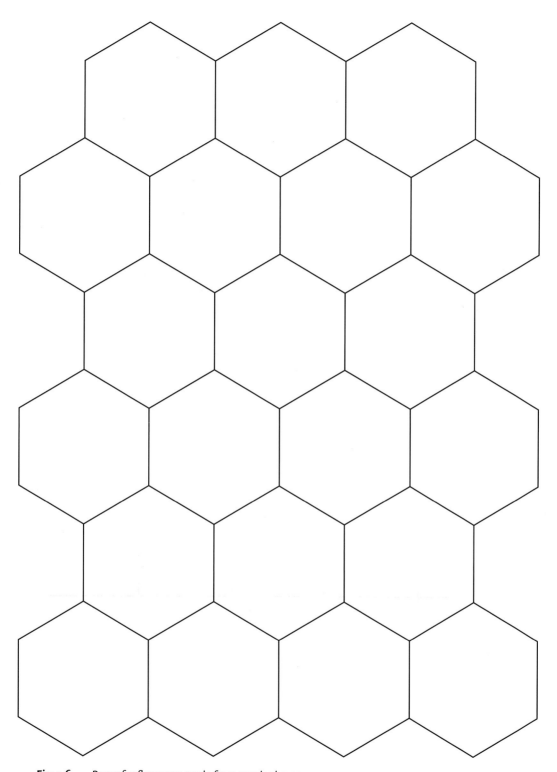

Fig. 1.6   Paper for flexagons made from regular hexagons.

Fig. 1.7   Paper for flexagons made from regular pentagons.

numbers, as described in Section 1.1 for the trihexaflexagon. A flexagon which is an untwisted band does not have an enantiomorph. Paper models of flexagons can be conveniently kept in transparent A5 (210 × 149 mm) size pockets held in A5 ring binders.

There are numerous published nets for various types of flexagon. Many of these are not in a form suitable for photocopying. Nets for flexagons made from equilateral triangles, squares or regular hexagons can conveniently be redrawn on photocopies of Fig. 1.4, 1.5 or 1.6. Where a net overlaps, or it is too long for these figures, it may be assembled from two or more parts. Nets for flexagons made from regular pentagons may be assembled from parts of strips photocopied from Fig. 1.7.

1.5 Decoration of faces

The appearance of paper models of flexagons can be improved by colouring and decorating the faces. Thus for example the faces of the trihexaflexagon numbered 1, 2 and 3 could be coloured red, yellow and blue respectively. The colours of visible faces would then change as it was flexed. Numerous decorative schemes have been used on various types of flexagon (Chapman 1961, Conrad 1960, Conrad and Hartline 1962, Gardner 1966, Johnson 1974, McIntosh 2000c, McIntosh 2000g, Mitchell 1999, Neale 1999).

Some of the decorative schemes exploit symmetries of flexagons both to create an attractive appearance and to create puzzles. There are books (Mitchell 1999, Pedersen and Pedersen 1973) that include decorated cut out nets for several types of flexagon. These have the disadvantage that making up the flexagons destroys the book. It is of course possible to make attractive and more durable models using other materials. At the time of writing an attractive model trihexaflexagon made from wood veneer triangles was available commercially. The triangles were glued to both sides of a strip of black cloth. Narrow cloth strips were left between adjacent triangles to form hinges.

2 Early history of flexagons

Flexagons are a twentieth century discovery. Arthur H. Stone, a postgraduate student at Princeton University in America, discovered them in 1939 while folding strips of paper. A Flexagon Committee was set up at Princeton University to investigate flexagons. In 1940 two members of the committee worked out a mathematical theory of flexagons but this was never published. The committee disbanded in 1941 following America's entry into the Second World War and for some time there was little visible activity. Interest revived in the late 1950s through the publication of two articles on flexagons in Scientific American and the publication of the first paper on flexagons. The subject can be said to have reached maturity with the issue in 1962 of a comprehensive report on flexagons. This report is a key reference, but it was not published in a form which reached a wide audience.

2.1 The discovery of flexagons

In terms of the history of mathematics, which goes back thousands of years (Cromwell 1997, Hirsch 1997, Sawyer 1943) flexagons are a relatively recent, twentieth century, discovery. Their early history is quite well documented as background information in publications on flexagons (Conrad and Hartline 1962, Cundy and Rollett 1981, Gardner 1965, Gardner 1966, Gardner 1988, Hilton et al. 1997, Johnson 1974, Kenneway 1987, McIntosh 2000a, McLean 1979, Madachy 1968, Mitchell 1999, Oakley and Wisner 1957).

Hexaflexagons were the first variety of flexagon to be discovered and investigated. They were discovered in the autumn of 1939 by Arthur H. Stone. He was a 23 year old English graduate student of mathematics at Princeton University in America. Stone had trimmed an inch (25 mm) from the edge of his American sheets of paper to make them fit his English binder. For amusement he began to fold the trimmed off strips of paper in various ways. One of the structures he made turned out to be the trihexaflexagon (Section 1.1).

The trihexaflexagon has three faces. On the next day Stone confirmed his belief, arrived at by overnight thoughts, that a more complicated paper model, with six faces instead of three, was possible. He found the structures so interesting that he showed his paper models of hexaflexagons to friends in the graduate school. Soon, numerous hexaflexagons were appearing at lunch and dinner tables.

2.2   The Flexagon Committee

A 'Flexagon Committee' was set up at Princeton University to investigate the properties of hexaflexagons. The members were Stone; Bryant Tuckerman, another graduate student of mathematics; Richard P. Feynman, a graduate student in physics; and John W. Tukey, a young mathematics lecturer. All four members subsequently had distinguished careers, with Feynman becoming a Nobel Laureate. The diagrams Feynman devised for the analysis of hexaflexagons were forerunners to the well known Feynman Diagrams in modern atomic physics. The name ' hexaflexagon' was coined. It was derived from 'hexa' for the hexagonal form, and 'flexagon' for the ability to flex. The name trihexaflexagon was derived from 'tri' for its three faces. Stone's elegant second structure was called a hexahexaflexagon (from its six faces).

In 1940 Tukey and Feynman worked out a complete mathematical theory of hexaflexagons. It is surprising that Feynman does not mention flexagons in his light hearted memoirs (Feynman 1989). Indeed, given the simplicity of the trihexaflexagon it is surprising that there do not appear to be any earlier references to it. The theory included a description of exactly how to construct a hexaflexagon of any desired type. The theory was never published, though parts of it have been rediscovered independently by other investigators. Among the early investigators was Tuckerman's father, the distinguished physicist Louis B. Tuckerman.

The committee, notably Tukey, also investigated square flexagons. They spent a lot of time analysing them, but did not succeed in developing a comprehensive theory that would cover all the numerous possible types. The committee were not the first to discover the simple square flexagon described in Section 1.3. In the modified form, with an extra row of squares as shown in Fig. 2.1, it has been used for centuries as the basis for a double acting hinge. This modified form is also used as the basis for a 'conjuror's wallet' which can make bank notes disappear and reappear (Gardner 1966, Laithwaite 1980).

Following America's entry into the Second World War in 1941 the Flexagon Committee disbanded and for some time there was little visible interest in flexagons.

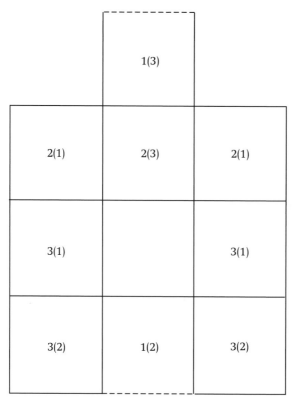

Fig. 2.1   Net for a paper model of a double acting hinge. Cut out the hole in the middle. Crease along the horizontal lines. Join at the dashed lines.

2.3   Revival of interest

In the late 1950s interest in flexagons revived through three main influences. Firstly, through Tuckerman senior, who had demonstrated items of flexagon theory to winners of a science talent contest each year for several years. Secondly, through Martin Gardner, who told the story of the discovery of flexagons, together with construction instructions, in two articles in *Scientific American* in 1956 and 1957. These articles later reached a much wider audience when reprinted as chapters in two books (Gardner 1965, Gardner 1966). Finally, through the publication of some of the theory of hexaflexagons by Oakley and Wisner (1957). This was the first publication on flexagons at the serious mathematical level.

The members of the original Flexagon Committee laid the groundwork of flexagon theory. The subject can be said to have reached maturity with the issue of a comprehensive report, at the serious mathematics level, by Conrad and Hartline (1962). This report describes in detail the theory of flexagons made from a wide range of polygons. The report is a key reference, but it was not published in a form which reached a wide audience. It contains much material that has subsequently been rediscovered by a number of people. At the time of writing, through the efforts of Harold V. McIntosh, the report could be downloaded from the Internet (http://delta.cs.cinvestav.mx/~mcintosh/oldweb/new.html).

3 Geometry of flexagons

In general the main characteristic feature of a flexagon is that it has the appearance of a polygon which may be flexed in order to display pairs of faces, around a cycle, in cyclic order. Another characteristic feature is that faces of individual polygons, known as leaves, which make up a face of a flexagon, rotate in the sense that different vertices move to the centre of a main position as a flexagon is flexed from one main position to another. The visible leaves are actually folded piles of leaves, called pats. Sometimes pats are single leaves. Alternate pats have the same structure. A pair of adjacent pats is a 'sector'.

A convenient mathematical framework for the analysis of flexagons is presented in this chapter, together with explanations of special technical terms. A straightforward geometric approach, without equations, is used. Geometric descriptions are used for three main purposes: firstly, to map the dynamic behaviour of flexagons, secondly to analyse their structure, and thirdly as the basis of recipes for the construction of flexagons of any desired type. Recipes are described later in the book.

There is a fundamental difficulty in mapping the dynamic behaviour of flexagons. This is that in general their structure and dynamic behaviour are so complicated that there has to be a compromise between including relevant detail and keeping a map reasonably easy to follow. The terminology used in the analysis of the structure of flexagons is mostly based on that used by previous authors, but to keep the text concise it was found that some new terms are needed. Maps and terminology are illustrated by material which properly belongs to later chapters. Specialised terms are listed in the subject index so that definitions can be easily located.

In geometry a distinction is often made between a mathematical ideal object and an imperfect physical model of the object. An ideal flexagon consists of a band of identical, rigid, flat leaves of zero thickness, which are hinged together at common sides. Fortunately, paper models of many types of flexagon do approximate closely to the mathematical ideal, and it isn't usually necessary to make a distinction between an ideal flexagon and the corresponding paper model. Paper isn't rigid. It also has nonzero thickness, which can interfere with the manipulation of a flexagon. It is sometimes regarded as permissible to bend the leaves during flexing.

3.1   Flexagon characteristics

It appears to be generally accepted that the main characteristic feature of a flexagon is that in a main position it has the appearance
of a polygon which may be flexed in order to display pairs of faces,
around a cycle, in cyclic order. There are, however, some flexagons where
it is only possible to traverse part of a cycle. In some, more complex,
flexagons it is possible to traverse two or more separate cycles, and to
traverse between these separate cycles. Another characteristic feature of
flexagons is that faces of individual leaves, which make up a face of a
flexagon, rotate in the sense that different vertices move to the centre
of a main position as a flexagon is flexed from one main position to another. This feature has been exploited aesthetically (Conrad and Hartline
1962, Gardner 1965, Gardner 1988, Hilton et al. 1997, Johnson 1974,
Kenneway 1987, Laithwaite 1980, Mitchell 1999, Pedersen and Pedersen
1973).

In geometry a distinction is often made between a mathematical ideal
object and an imperfect physical model of the object. For example, a
line is defined as having zero width whereas any real line drawn on a
piece of paper must have a nonzero width. Fortunately, paper models
of many types of flexagon do approximate closely to the mathematical
ideal, and it isn't usually necessary to make a distinction between an
ideal flexagon and the corresponding paper model. Ideally, the leaves of
a flexagon should be rigid and should also have zero thickness. Paper
isn't rigid. It also has nonzero thickness, which can interfere with the
manipulation of a flexagon. Good quality 80 g/m^2 printer paper is a good
compromise between rigidity and thickness. It is sometimes regarded as
permissible to bend the leaves during flexing.

An ideal flexagon consists of a band of identical (congruent), rigid,
flat leaves, which are hinged together at common sides. An ideal leaf is
regarded as consisting of its one dimensional sides plus its two dimensional interior (Cromwell 1997). Here, 'hinged' means that the dihedral
angle between the two planes containing two hinged leaves may vary
between 0° and 360° without constraint. The dihedral angle (Coxeter
1963) is the angle on a section which cuts both planes at 90°. As an
ideal leaf is of zero thickness the pats, which are either single leaves
or folded piles of leaves, are also of zero thickness. This leads to the
conceptual difficulty of deciding exactly what is meant by 'upwards' or
'downwards' through a pat. However, in a paper model of a flexagon
the leaves do have nonzero thickness and the conceptual difficulty is
avoided.

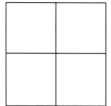

Fig. 3.1   Appearance of a main position of a square flexagon.

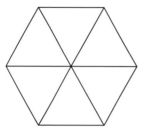

Fig. 3.2   Appearance of a main position of a hexaflexagon.

3.2   Appearance of main positions

In investigating possible varieties of flexagon Chapman (1961) considered possible arrangements of an even number, $2n$, of regular convex polygons, with s sides, about a point in a plane. He pointed out that there are two solutions. These correspond to main positions of square flexagons ($n = 2$, $s = 4$) and hexaflexagons ($n = 3$, $s = 3$). The appearances of these main positions are shown in Figs. 3.1 and 3.2. He concluded that they are the only possible flexagon varieties. As the main positions in the figures, by definition, lie in a plane it is convenient to call them 'flat' main positions. Also, by definition, the sum of leaf vertex angles at the centres of the main positions is $360°$.

Other types of main position are possible if Chapman's approach is relaxed. If the sum of the leaf vertex angles at the centre of a main position is greater than $360°$ then a main position is not flat and its outline is a skew polygon. As an example Fig. 3.3 shows the appearance of a main position of a pentagon flexagon. Such a main position may be called a 'skew main position'. If the sum of the leaf vertex angles at the centre of a main position is less than $360°$ this results in a 'slant main position'. The outline is a flat polygon. As an example Fig. 3.4 shows the appearance of a main position of a triangle flexagon. With some varieties of flexagon there is more than one type of main position appearance and it is necessary to distinguish between 'principal' and 'subsidiary' main positions

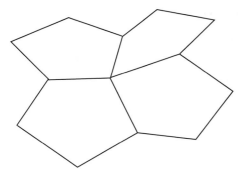

Fig. 3.3   Appearance of a principal main position of a pentagon flexagon.

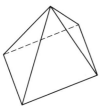

Fig. 3.4   Appearance of a main position of a triangle flexagon.

and cycles. This is not needed for square flexagons, hexaflexagons and triangle flexagons. Pentagon flexagons have two types of main position appearance. As noted in the caption Fig. 3.3 shows the appearance of a principal main position of a pentagon flexagon.

3.3   Flexagons as linkages

From a mechanical engineering viewpoint an ideal flexagon is a three dimensional linkage. The formal definition of a linkage (Macmillan 1950) is that it is an assembly of coupled rigid bodies (links) whose freedom of movement is restricted, after the fixture of one link in space, by the constraint imposed by their couplings. The number of degrees of freedom possessed by a linkage is usually defined as the number of independent parameters needed to completely determine its configuration. In any practical linkage the number of links is finite so the number of degrees of freedom is also finite. In other words the links can only follow a finite number of paths relative to each other. In an ideal flexagon the links are the identical, rigid leaves each of which is coupled to two neighbouring leaves by hinges along common sides. It is therefore what is known as a hinged linkage.

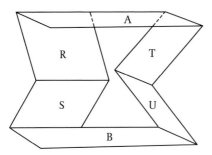

Fig. 3.5   Sarrut's parallel motion.

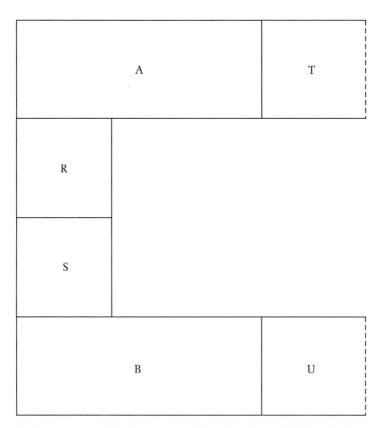

Fig. 3.6   Net for a paper model of Sarrut's parallel motion. Crease, fold into the form shown in Fig. 3.5 and join the ends.

3.3.1   Sarrut's parallel motion

Sarrut's (sometimes Sarrus's) parallel motion is an example of a hinged linkage which illustrates some of their possible features. It was discovered in 1853 (Dunkerley 1910). It is shown in Fig. 3.5, as usually illustrated. The net for a paper model is shown in Fig. 3.6. The point of Sarrut's parallel

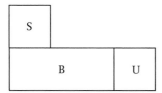

Fig. 3.7   Sarrut's parallel motion collapsed to lie flat.

motion is to ensure that the moving link A remains parallel to the fixed link B, which is of the same size and shape. To achieve this four connecting pieces of equal length, R, S, T and U, are used. The pieces A, R, S and B are hinged in sequence by three parallel hinges. The pieces A, T, U, and B are similarly hinged, but these hinges have a different orientation. A textbook description of Sarrut's parallel motion (Dunkerley 1910) states that it has one degree of freedom. To see this for the partially extended configuration (Fig. 3.5) choose the distance between pieces A and B as an independent parameter. If this distance is fixed then the parallel motion becomes rigid. Hence the configuration is completely determined so there is one degree of freedom.

When Sarrut's parallel motion is fully extended the situation is more complicated because the freedoms are infinitesimally greater than one and change as soon as link A starts to move towards link B. In this situation it is difficult to select appropriate independent parameters. Tabs RS and TU can each fold either inwards or outwards as link A starts to move towards link B. Hence there are four possible ways in which the parallel motion can start to move. Strictly speaking the situation may be described by two parameters, one for each tab, both of which may be either positive, for one direction of folding, or negative for the other. In practice it is easier to regard the number of degrees of freedom as the number of ways (four) in which the parallel motion can move from the fully extended position.

If the tabs fold outwards from the fully extended position and the linkage is fully collapsed to lie flat then the configuration shown in Fig. 3.7 is reached. Between the two extreme positions there is one degree of freedom, as in the textbook definition given above. In the fully collapsed position there are two additional degrees of freedom in which the tabs RS and TU can be rotated about hinges, giving a total of three. However, if one or both of these tabs are rotated out of plane then it is no longer possible to move piece A relative to piece B and there are two degrees of freedom. If the tabs fold inwards from the fully extended position they become trapped when the linkage is fully collapsed and there is then only one degree of freedom.

3.3.2   Degrees of freedom of an ideal flexagon

Changes in the number of degrees of freedom occur during the flexing of a flexagon, but they can be difficult to enumerate. The changes in the number of degrees of freedom of a flexagon as it is flexed between configurations are an important characteristic which dominate its dynamic behaviour. Their effects are easily observed in paper models.

As an example, consider the simple square flexagon described in Section 1.3. At its intermediate position it has the appearance of two squares hinged together, and there are three degrees of freedom. Two of these correspond to rotation about hinges which lead to the two main positions, and the third to rotation about the set of hinges connecting the two pats. During rotation to a main position there is one degree of freedom. At a main position (Fig. 3.1) there are two degrees of freedom. These correspond to rotations about the two pairs of hinges connecting the four pats.

3.4   Mapping flexagons

There is a fundamental difficulty in mapping the dynamic behaviour of flexagons. This is that their structure and dynamic behaviour are so complicated, for all except the simplest flexagons, that there has to be a compromise between including relevant detail and keeping a map reasonably easy to follow. Several different types of maps have been used (Chapman 1961, Conrad 1960, Conrad and Hartline 1962, Gardner 1965, Gardner 1966, Hilton and Pedersen 1994, Hilton et al. 1997, Liebeck 1964, McIntosh 2000a, McIntosh 2000c, McIntosh 2000d, McIntosh 2000e, McIntosh 2000f, McIntosh 2000g, McIntosh 2000h, McLean 1979, Madachy 1968, Mitchell 1999, Wheeler (1958).

3.4.1   Full map of the irregular single cycle square flexagon

The irregular single cycle square flexagon is used both to illustrate the complexity of the dynamic behaviour of flexagons and to show how a 'full map' (Fig. 3.8) can be used to describe and interpret this behaviour. The net for the irregular single cycle square flexagon is shown in Fig. 3.9. This flexagon has four faces and four main positions, which can be visited in cyclic order. It is an untwisted band so there is no enantiomorphic (mirror image) form.

As assembled the flexagon is in the main position shown as the top left diagram in Fig. 3.8. Faces numbers 1 and 2 are visible and this is main position 2(1). The first number in the code identifying a main position

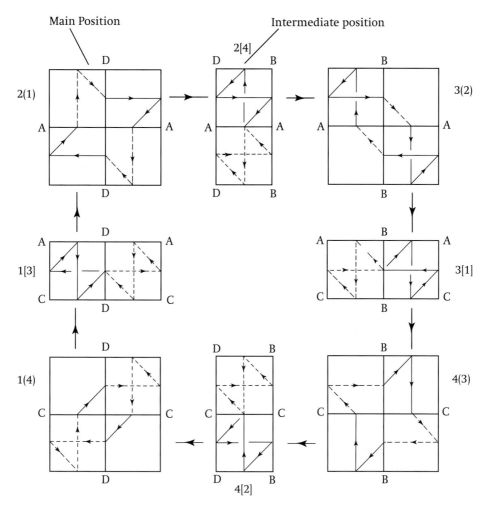

Fig. 3.8 Full map of the irregular single cycle square flexagon.

indicates the numbers visible on the upper face and the second number, in round brackets, those visible on the lower face. If the flexagon were turned over so that face number 1 were uppermost then the code would be 1(2). It is sometimes not necessary to indicate which face of a flexagon is uppermost and the codes 1(2) and 2(1) are then synonymous. The main positions are flat, and have twofold rotational symmetry about a vertical axis through the centre. A main position therefore consists of two identical 'sectors'. There are two adjacent pats in a sector. In general in a main position a flexagon has $2n$ pats and n sectors.

The lines A-A and D-D on the top left diagram identify the hinges on the square flexagon, as marked on the net, and hence its orientation. Rotating a main position of the square flexagon through 180° about a vertical axis makes no difference because it has twofold rotational symmetry.

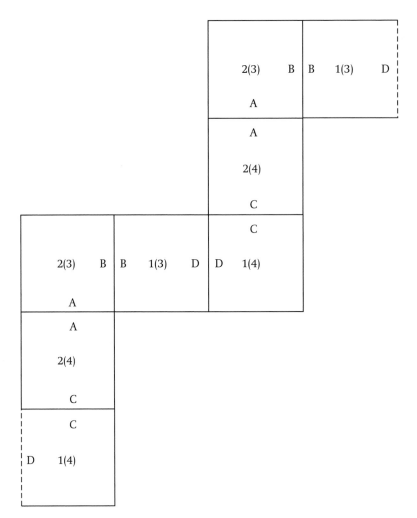

Fig. 3.9   Net for the irregular single cycle square flexagon.

The orientations shown in Fig. 3.8 are arbitrary, and have been chosen to assist visualisation of the dynamic behaviour of the flexagon. In a main position a square flexagon has two degrees of freedom in that rotation can take place about two different hinge lines. In main position 2(1) this is either hinge line A-A or hinge line D-D.

The full map shows what happens when the square flexagon is flexed around its cycle using the pinch flex. The description below is for a traverse in the direction of the arrows in Fig. 3.8. Starting from main position 2(1), fold the square flexagon in two, along hinge line D-D, to reach intermediate position 2[4]. Keep the hinge line uppermost and ensure that the leaves numbered 1 are concealed. An intermediate position has the appearance of two squares, joined at a common side, and is flat. It has

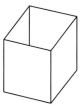

Fig. 3.10 Square flexagon box position.

twofold rotational symmetry about the central hinges, for example hinge line A-A in intermediate position 2[4]. The top centre diagram in the figure shows the appearance from one side. Only one number is visible, and this is the first number shown in the code. Open the flexagon at the top and unfold about hinges B so as to reach main position 3(2). Continue flexing in cyclic order in order to traverse the complete cycle. If the flexagon is turned over, the cycle may be traversed in the reverse direction.

At a main position it is possible to fold a square flexagon in two in four different ways to reach four different intermediate positions. However, for the irregular single cycle square flexagon two of these do not lead to another main position. Such intermediate positions are not normally shown on a full map.

3.4.2 The box position

In general, degrees of freedom are the same as for the simple square flexagon (Subsection 3.3.2). However, there are four degrees of freedom at the intermediate positions. At intermediate position 2[4] (Fig. 3.8) the flexagon can be unfolded about hinge line B-B or D-D, or folded about hinge line A-A. It can also be opened up into an open ended box. This is called a 'box position' (Fig. 3.10). If this is done then the second number in the code for an intermediate position, in square brackets, becomes visible inside the box. The intermediate position codes indicate that flexing between intermediate positions 1[3] and 3[1], or between intermediate positions 2[4] and 4[2], turns the box position inside out.

3.4.3 Pat structure of the irregular single cycle square flexagon

In main positions 2(1) and 4(3) of the irregular single cycle square flexagon (Fig. 3.8) each pat is a folded pile of two leaves. However, in main positions 3(2) and 1(4) alternate pats are a single leaf and a folded pile of three leaves. A notation is needed to characterise the internal structure of the pats. To accomplish this line segments connecting midpoints of hinges have been included in the main and intermediate position diagrams. The convention used is that if a circuit of these line segments is traversed

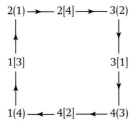

Fig. 3.11 Simplified map of the irregular single cycle square flexagon.

Fig. 3.12 Tuckerman diagram for the irregular single cycle square flexagon.

in the direction of the arrows, then a solid line indicates that a circuit passes upwards through a pat, whereas a dashed line indicates that the circuit passes downwards. A further convention is that a line segment is broken where it passes below another part of the circuit. When a pat is a single leaf then the leaf connects the top of one adjacent pat to the bottom of the other adjacent pat. The line segment type shows the sense of this connection.

If the distinguishing face numbers and hinge letters are ignored then it can be seen (Fig. 3.8) that there are two types of main position. Main positions 2(1) and 4(3) are of one type and are identical, although this is not immediately obvious from the figure. The identity becomes clear if one of the main positions is turned over and its circuit is redrawn. Main positions 3(2) and 1(4) are of another type and are an enantiomorphic (mirror image) pair. As there are two distinct types of main position it is appropriate to describe this particular square flexagon as irregular in its dynamic behaviour, hence its name.

3.4.4 Some other types of map

Some other types of map are used later in the book, and the irregular single cycle square flexagon is used below to illustrate these.

Fig. 3.11 is a 'simplified map' showing just the main and intermediate position codes, and the direction of traverse.

Fig. 3.12 is a further simplification. This is known as a Tuckerman diagram (Conrad and Hartline 1962, Gardner 1965, Hilton and Pedersen 1994). Tuckerman diagrams indicate what pairs of faces are displayed in

Fig. 3.13   Intermediate position map of the irregular single cycle square flexagon.

Fig. 3.14   Flexagon figure for the irregular single cycle square flexagon.

all the possible main positions of a flexagon, and also how to traverse from one main position to any of the others. Hence, only main position codes and the direction of traverse are shown.

Omitting the main position codes and the direction arrows results in a 'flexagon diagram', which is simply a square.

In an 'intermediate position map' (Fig. 3.13) the cycle is represented by a square. The numbered vertices represent intermediate positions, with each number corresponding to the number visible on the outside of an intermediate position. The sides represent main positions, with the numbers at the vertices corresponding to the numbers visible on a main position. For example the '1' at the top left hand vertex represents intermediate position 1[3] (Fig. 3.8) and the line at the top of the square represents main position 2(1).

3.5   Flexagon figures and symbols

A 'flexagon figure' may be used to characterise any single cycle flexagon in an economical manner. It consists of two polygons, taken as just their sides. The first is a circumscribing polygon which is the same type of polygon as the leaves used to construct the flexagon. The second part is an inscribed polygon whose vertices lie on midpoints of sides of the circumscribing polygon. This inscribed polygon is an invariant characterising the dynamic behaviour of the flexagon, and is obtained from the part circuits shown in the intermediate positions on a full map.

The part circuit for each pat in the intermediate positions of the irregular single cycle square flexagon is a crossed quadrilateral (Fig. 3.8) and is characteristic of this particular square flexagon. It is used as the inscribed polygon in the flexagon figure, but without the direction arrows, as shown in Fig. 3.14. The circumscribing polygon is a square.

Fig. 3.15   Flexagon figure for the trihexaflexagon.

Fig. 3.16   Regular star pentagon.

The trihexaflexagon (Section 1.1) is a regular flexagon in that all the main positions are identical. Its flexagon figure is an equilateral triangle inscribed in another equilateral triangle (Fig. 3.15).

A flexagon figure contains enough information to construct the corresponding flexagon. There is a unique solution for the irregular single cycle square flexagon, but the trihexaflexagon exists in two enantiomorphic forms so there are two solutions.

Characterising regular polygons through the use of Schläfli symbols is well established (Coxeter 1963, Wenninger 1971). The Schläfli symbol for a regular convex polygon is $\{s\}$, where s is the number of sides. It completely defines a regular convex polygon, except for its size. For example the Schläfli symbol for an equilateral triangle is $\{3\}$ and for a square it is $\{4\}$. The Schläfli symbol for a regular star pentagon (Fig. 3.16) is $\{5/2\}$. The 5 indicates that there are five sides and the 2 indicates that the interior is covered twice. This means that a line drawn from the centre of the regular star pentagon to its exterior crosses two sides.

A regular flexagon is completely defined by its 'flexagon symbol' $\langle s, c \rangle$ where $\{s\}$ is the Schläfli symbol for the leaves used to construct the flexagon, and $\{c\}$ is the Schläfli symbol for the inscribed polygon of the flexagon figure. For example the flexagon symbol for the trihexaflexagon is $\langle 3, 3 \rangle$. In this example, as is often the case, $s = c$, but there are flexagons where $s \neq c$.

4 Hexaflexagons

Hexaflexagons were the first variety of flexagon to be discovered and they have been analysed in the most detail. The leaves of a hexaflexagon are equilateral triangles. In appearance a main position of a hexaflexagon is flat and consists of six leaves, each with a vertex at the centre so there are six pats and three sectors. The outline is a regular hexagon. All hexaflexagons are twisted bands and hence exist as enantiomorphic (mirror image) pairs. Enantiomorphs are not usually regarded as distinct types. The handedness of a hexaflexagon is only mentioned when this is needed for clarity.

In some ways hexaflexagons are the simplest type of flexagon. There are only one possible type of cycle and one possible type of link between cycles and normally only the pinch flex is used. On the other hand hexaflexagons do have a large number of degrees of freedom. In consequence with most types of hexaflexagons spurious transformations may occur during flexing, and it is possible to get some types badly tangled.

The trihexaflexagon has three faces and only one cycle. Since there are three main positions this is called a '3-cycle'. Its decorative possibilities are illustrated by a transformation from a happy face to a sad face. Multicycle hexaflexagons have been extensively analysed and their dynamic behaviour is well understood. A 'Tuckerman traverse' is a systematic method of traversing a multicycle hexaflexagon which guarantees that all possible main positions are visited. Numerous nets have been published, and some examples are given. The number of distinct types of hexaflexagon with the same number of faces increases rapidly with the number of faces and some data are tabulated.

There has been much interest in the design of nets for specific types of hexaflexagon whose dynamic behaviour is known, and some useful generalisations are possible. Starting from a known design, practical methods based on paper models make it possible to add or delete cycles and faces, and hence design hexaflexagons with desired dynamic behaviours. These methods are described in detail.

There are six possible arrangements of the leaves which make up a face of a hexaflexagon. These are illustrated by a specially designed seven faced hexaflexagon which includes all six possible arrangements of one of its faces. These possible arrangements are then used as the basis of a decorated puzzle version of one of the possible six faced hexaflexagons, the hexahexaflexagon.

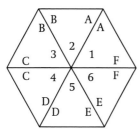

Fig. 4.1   Null hexaflexagon main position and net. Crease the lines between triangles to form hinges.

4.1   Degrees of freedom of hexaflexagons

Hexaflexagons have a large number of degrees of freedom. In consequence with most types of hexaflexagons spurious transformations may occur if the pinch flex is not carried out correctly. It is possible to get some types badly tangled, with differently numbered leaves appearing on the same face. The need to avoid spurious transformations is the reason why some authors give very detailed instructions on the manipulation of paper models of hexaflexagons. Such instructions include the implicit requirement that threefold rotational symmetry be maintained during pinch flexing, as shown in Fig. 1.2. This artificially limits the number of degrees of freedom, and avoids spurious transformation.

4.1.1   The null hexaflexagon

The degrees of freedom of hexaflexagons may initially be considered in terms of the degrees of freedom of the null hexaflexagon. The null flexagon has two faces. Its one main position is its own net, so assembly is not required (Fig. 4.1). Degrees of freedom are counted in terms of the number of ways the full flexagon may be folded rather than in terms of parameters which may be either positive or negative (cf. Section 3.3.1).

In the main position it has 31 degrees of freedom, as follows. It can be rotated about a pair of collinear hinges in 3 different ways. A single pinch can be used to press two adjacent triangles together in 24 different ways. Triangles 1 and 2 can be pressed together with valley folds at hinges B and F, and peak folds at hinges A, C, D and E, or with valley folds at hinges B, C, D, E and F, and a peak fold at hinge A, and in 2 more ways with peak and valley folds interchanged; there are 6 different pairs of triangles which can be pinched together. Three pairs of triangles can be pinched together, using the first stage of the pinch flex (Section 1.2), in 4 different ways. For example triangles 1 and 2, 3 and 4, and 5 and 6 can be pressed together with peak folds at hinges A, C and E, and valley folds at hinges B, D and F.

If in the main position the null hexaflexagon is folded in two about a pair of collinear hinges, say A and D, there are then 5 degrees of freedom, as follows. It can be unfolded about hinges A and D or rotated about either of the other two pairs of hinges. It is also possible to keep two triangles, say 1 and 2, together and open the other four into a pyramid without a base. This pyramid can then be closed in the opposite direction and the triangles then rotated to reach an intermediate position. There are therefore two different ways of flexing between the main and an intermediate position. This shows that the requirement to maintain threefold rotational symmetry is a sufficient, but not a necessary, condition for correct flexing.

In an intermediate position there are 3 degrees of freedom, as follows. There are 2 degrees of freedom formed by rotations about the hinges at the common edge. The third is to return the null hexaflexagon to its main position by half a pinch flex. In any other type of hexaflexagon there are two types of intermediate position. There is one in which it is only possible to return to one main position by half a pinch flex in one way so that there are 3 degrees of freedom, as for the null hexaflexagon. Such intermediate positions are not normally shown on full maps. In the other type of intermediate position it is possible to use half a pinch flex in two different ways to reach two different main positions, and there are 4 degrees of freedom. If threefold rotational symmetry be imposed then there are respectively 1 and 2 degrees of freedom for the two types of main position.

4.2   The trihexaflexagon revisited

The trihexaflexagon has three faces and three main positions (Section 1.1), and it is the only possible type of single cycle hexaflexagon. It is a Möbius band with three half twists. A Möbius band (Gardner 1965, Gardner 1978, Pedersen and Pedersen 1973) has only one side (face) and one edge. Fig. 4.2 shows a trihexaflexagon main position sketched so as illustrate this feature. Since it is a twisted band the trihexaflexagon exists as an enantiomorphic (mirror image) pair. The net for the left handed enantiomorph is shown in Fig. 4.3. This is the same as the net shown in Fig. 1.1 except that hinge letters have been added.

The dynamic properties of the trihexaflexagon are shown in the full map (Fig. 4.4). The cycle has three main positions so it is called a '3-cycle'. Each main position is flat and has threefold rotational symmetry about a vertical axis through the centre. It consists of three sectors. If the distinguishing face numbers and hinge letters are ignored then all three main positions are identical, and the trihexaflexagon is regular, as noted

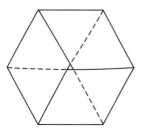

Fig. 4.2   Trihexaflexagon main position as a Möbius band with three half twists.

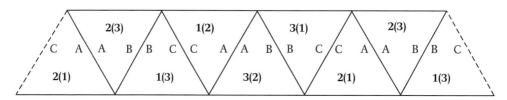

Fig. 4.3   Net for the trihexaflexagon.

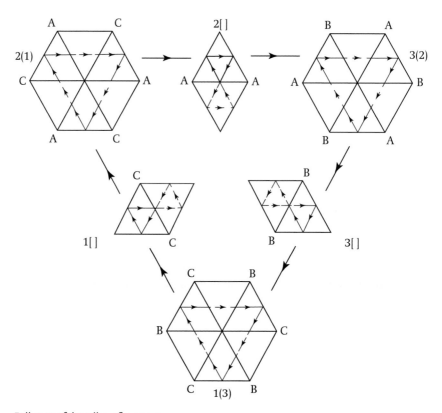

Fig. 4.4   Full map of the trihexaflexagon.

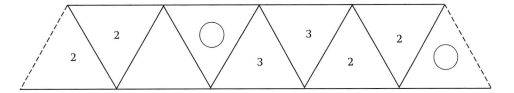

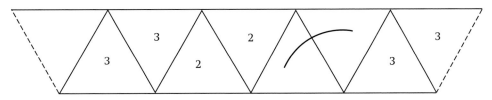

Fig. 4.5   Net for the trihexaflexagon. Face 1 decorated. Both sides of net shown.

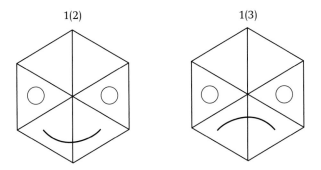

Fig. 4.6   Appearances of face 1 of the decorated trihexaflexagon.

in Section 3.5. The hinge letters give an indication of the orientation of
the individual triangles visible on main position faces. In a main position
alternate pats are single leaves and folded piles of two leaves. Each intermediate position consists of three pats hinged together at common sides
with an angle of 120° between pats. Only two pats are shown on the map,
folded flat. The third pat may be imagined as extending vertically down-
ward. An intermediate position cannot be opened up into a box. It can be
flexed to a main position in two different ways so there are 4 degrees of
freedom. The intermediate positions show that the flexagon figure is an
equilateral triangle inscribed in another equilateral triangle (Fig. 3.15),
so the flexagon symbol is $\langle 3, 3 \rangle$.

The decorative possibilities of the trihexaflexagon are illustrated by
the net shown in Fig. 4.5. Face 1 is now decorated, instead of being numbered as in Fig. 4.3. The two possible appearances of face 1 of the decorated trihexaflexagon are shown in Fig. 4.6. In main position 1(2) a happy

face is displayed, but in main position 1(3) this becomes a sad face. This transformation is due to Alan Phillips (Conrad and Hartline 1962).

4.3   Multicycle hexaflexagons

In one way hexaflexagons are the simplest variety of flexagon in that only one type of cycle is possible and this has only three main positions. Multicycle hexaflexagons have been extensively analysed and numerous nets have been published, together with assembly instructions (Conrad 1960, Conrad and Hartline 1962, Cundy and Rollett 1981, Gardner 1965, Gardner 1988, Hilton and Pedersen 1994, Hilton et al. 1997, Johnson 1974, Kenneway 1987, Laithwaite 1980, Liebeck 1964, McIntosh 2000h, McLean 1979, Madachy 1968, Maunsell 1954, Mitchell 1999, Oakley and Wisner 1957, Pedersen and Pedersen 1973, Wheeler 1958).

4.3.1   Numbers of distinct types

At the serious mathematics level there has been interest in determining the numbers of distinct types of hexaflexagon with the same number of faces (Conrad 1960, Conrad and Hartline 1962, Gardner 1976, Oakley and Wisner 1957, O'Reilly 1976, Sloane and Plouffe 1995). Numbers of distinct types of hexaflexagon with up to 18 faces are shown in Table 4.1. The number of cycles is 2 less than the number of faces. Beyond 10 faces the number of distinct types increases rapidly and making a complete set becomes impractical. Note that data in Oakley and Wisner (1957) are incorrect.

4.3.2   The four faced hexaflexagon

There is only one type of hexaflexagon with four faces. Its net is shown in Fig. 4.7. There are two separate cycles which may be traversed, with a common main position. This type of link between cycles is a 'main position link'. It is the only possible type of link for hexaflexagons. A closed loop of cycles is not possible for main position links (Conrad and Hartline 1962, Wheeler 1958). The Tuckerman diagram for the four faced hexaflexagon is shown in Fig. 4.8 and the flexagon diagram in Fig. 4.9.

To traverse a cycle the hexaflexagon is rotated by 60° between pinch flexes. To traverse from one cycle to another the flexagon is not rotated. At main position 1(2) there is a choice of whether or not to rotate between flexes. At the other main positions the hexaflexagon must be rotated between flexes. In main position 1(2) all the pats are folded piles of two leaves (Fig. 4.10). In the other main positions alternate pats are single leaves and folded piles of three leaves as in Fig. 4.11, which shows main position 3(1). In this main position the pat structure is complex with the

Table 4.1 *Numbers of distinct types of hexaflexagon.*

Number of faces	Number of types
3	1
4	1
5	1
6	3
7	4
8	12
9	27
10	82
11	228
12	733
13	2 282
14	7 528
15	24 834
16	83 898
17	285 357
18	983 244

Conrad (1960), Conrad and Hartline (1962)

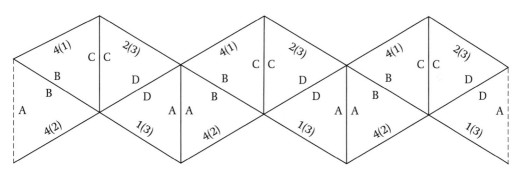

Fig. 4.7   Net for the four faced hexaflexagon.

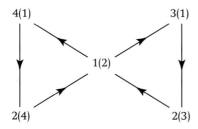

Fig. 4.8   Tuckerman diagram for the four faced hexaflexagon.

Fig. 4.9   Flexagon diagram for the four faced hexaflexagon.

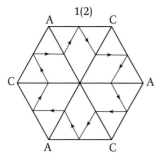

Fig. 4.10   Diagram of main position 1(2) of the four faced hexaflexagon.

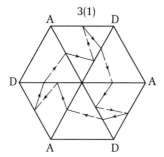

Fig. 4.11   Diagram of main position 3(1) of the four faced hexaflexagon.

circuit passing both upwards and downwards through the pat. Coincident parts of the circuit have been separated for clarity. Leaf faces numbered 4 are locked together in pairs in the pats in much the same way that tabs are locked in a fully collapsed configuration of Sarrut's parallel motion (Subsection 3.3.1). However, this is not obvious on casual examination of the four faced hexaflexagon.

4.3.3   The Tuckerman traverse

In a Tuckerman traverse a hexaflexagon is only rotated between pinch flexes when this is necessary for flexing to be continued. This guarantees that all possible main positions are visited. It is a useful method of finding the Tuckerman diagram for a hexaflexagon whose dynamic behaviour is unknown. If the hexaflexagon is turned over then the direction of traverse is reversed. Two conventions are possible when laying out a Tuckerman

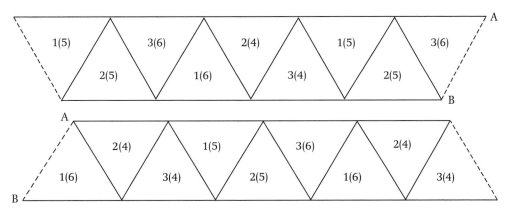

Fig. 4.12 Net for the hexahexaflexagon. Join the two parts of the net at A-B.

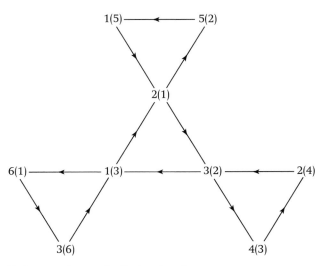

Fig. 4.13 Tuckerman diagram for the hexahexaflexagon.

diagram. In the convention used in Fig. 4.8 the cycles are oriented so that if a hexaflexagon is not rotated at a main position then the corresponding direction arrows are in a straight line.

4.3.4 The hexahexaflexagon

The net, Tuckerman diagram and flexagon diagram for one of the three possible types of six faced hexaflexagon are shown in Figs. 4.12, 4.13 and 4.14. Its dynamic behaviour has a high degree of symmetry (Fig. 4.13). Its elegance makes it a classical flexagon and it has been given a special name, the hexahexaflexagon (Section 2.2). At the time of writing an attractive wood model, with cloth hinges, was available commercially. Six

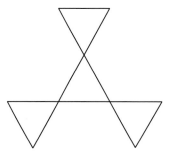

Fig. 4.14   Flexagon diagram for the hexahexaflexagon.

faces is close to the maximum number of faces for which a paper model
of a hexaflexagon can be conveniently manipulated.

In main positions 1(3), 2(1) and 3(2) (Fig. 4.13) alternate pats are folded
piles of two and four leaves. In the other main positions alternate pats
are single leaves and folded piles of five leaves. It is always found that at
a main position of a hexaflexagon where there is no link to another cycle
alternate pats are single leaves. Finding all nine possible main positions
of a hexahexaflexagon is quite a challenge for the uninitiated, especially
if the faces are decorated rather than numbered.

The net for the hexahexaflexagon is a straight strip of triangles
(Fig. 4.12). This is always true for a hexaflexagon whose flexagon diagram
has threefold rotational symmetry, as in Fig. 4.14. The nets for the other
two types of six faced hexaflexagon are crooked strips.

4.4   Hexaflexagon design

There has been much interest in the design of nets, including appropriate face numbering schemes, for specific types of hexaflexagon whose
dynamic behaviour is known (Conrad 1960, Conrad and Hartline 1962,
Hilton and Pedersen 1994, Liebeck 1964, McIntosh 2000b, McIntosh 2000c,
Madachy 1968, Wheeler 1958).

Some useful generalisations are as follows. There are six leaves visible on each face of a hexaflexagon and each leaf in a net has two faces,
so it follows immediately that the number of leaves in the net is three
times the number of faces. A hexaflexagon has three sectors so the sequence of face numbers on a net appears three times. All hexaflexagons
are twisted bands, and therefore exist as enantiomorphic (mirror image)
pairs. Enantiomorphs are not regarded as distinct types in Table 4.1. The
number of half twists in a hexaflexagon is the same as the number of
faces. Bands with an even number of half twists have two faces and two
edges. Those with an odd number of half twists have one face and one

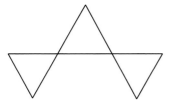

Fig. 4.15   Flexagon diagram for the five faced hexaflexagon.

edge, and are Möbius bands (Gardner 1965, Gardner 1978, Pedersen and Pedersen 1973).

At the serious mathematics level theoretical methods are used for net design, including the determination of face numbering schemes. Algorithms which have been developed, for example Liebeck (1964) and Madachy (1968), are based on some type of map which describes the dynamic behaviour of the target hexaflexagon. They always lead to two solutions, one for each enantiomorph. They may be incorporated in computer programs (McIntosh 2000b), but are tedious if done by hand. For hexaflexagons with small numbers of faces they have no advantage over practical methods using paper models, as described below.

4.4.1   Deleting faces

If the flexagon diagram for the target hexaflexagon can be formed by removing a triangle, or triangles, from the flexagon diagram for a known hexaflexagon then the net can be derived by deleting faces, one at a time, from the net for the known hexaflexagon. For example the flexagon diagram for the only possible five faced flexagon, shown in Fig. 4.15, can be obtained by removing one triangle from the flexagon diagram for the hexahexaflexagon (Fig. 4.14). The Tuckerman diagram for the hexahexaflexagon (Fig. 4.13) shows that one triangle can be removed by deleting any of the faces 4, 5 and 6. To delete one these faces from the net for the hexahexaflexagon (Fig. 4.12) glue together appropriately numbered faces of pairs of leaves in a paper model of the net. That is, to delete, say, face 4 glue together faces of leaves numbered 4. Alternatively, fasten the leaves together temporarily with paper clips and copy the net. Some renumbering of faces may be needed for a satisfactory numbering scheme. For each face removed one triangle is removed from a flexagon diagram and three leaves are removed from the corresponding net.

4.4.2   The synthetic method

In the synthetic method of hexaflexagon design two paper models of hexaflexagons are linked at main positions (Conrad 1960, Conrad and

Hartline 1962, Hilton and Pedersen 1994, Liebeck 1964, McIntosh 2000b, McIntosh 2000c, Madachy 1968, Wheeler 1958).

The two hexaflexagons linked must be of the same handedness. The flexagon diagrams are correspondingly linked. For example two trihexaflexagons (Section 4.2) may be linked to form the four faced hexaflexagon (Subsection 4.3.2). The procedure is as follows. Flex each hexaflexagon to the main position which is to be linked. This must be one in which alternate pats are single leaves. Cut the flexagon along the hinge lines between leaves, and remove the single leaves. Keep the remaining pats in their original relationship to one another. Mark the cut sides of leaves in the pats for future reference. Assemble the common main position of the hexaflexagons being linked by placing the remaining pats from each of the original hexaflexagons alternately. Join the marked side of the top leaf of each pat to the marked side at the bottom leaf of an adjacent pat. Renumbering of faces will usually be needed. In linking the two hexaflexagons six triangles and hence two faces are removed.

4.5   Hexaflexagon face arrangements

The arrangements of the leaves making up a face of a hexaflexagon change as it is flexed from one main position to another. There are six possible arrangements of the leaves on a face of a hexaflexagon. The changes are used both as the basis of puzzles (Gardner 1965, Laithwaite 1980, Mitchell 1999) and as the basis of decorative schemes (Conrad and Hartline 1962, Gardner 1965, Gardner 1988, Hilton et al. 1997, Johnson 1974, Kenneway 1987, Laithwaite 1980, Mitchell 1999, Pedersen and Pedersen 1973).

4.5.1   A seven faced hexaflexagon

The simplest hexaflexagon on which all six possible arrangements of a face occur is the seven faced hexaflexagon whose net is shown in Fig. 4.16. The Tuckerman diagram (Fig. 4.17) shows that face 1 is visible in six main positions. Face 1 is not numbered. Instead letters are used to identify the vertices of triangles and Roman numerals to identify individual triangles. The orientation of these letters and numerals is not regarded as having any significance. The changes in the arrangements of faces as the hexaflexagon is flexed from main position 1(2) to main position 1(7) are shown in Fig. 4.18. The letters at the centre of a face are in the sequence ABCABC, and adjacent leaves rotate in opposite directions from one main position to the next. The Roman numerals run alternately clockwise and anticlockwise during the sequence. In combination these changes mean that there are six possible arrangements, all of which are shown in the figure.

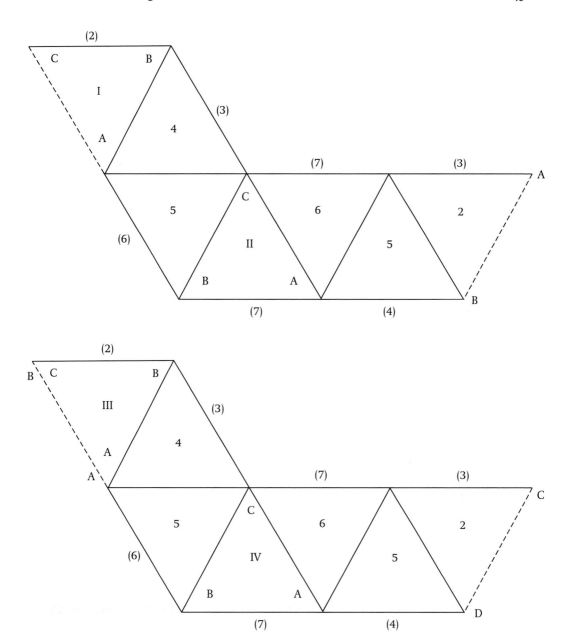

Fig. 4.16 Net for a seven faced hexaflexagon. Face 1 is not numbered. Write each number in brackets on the reverse of the adjacent triangle. Join the three parts of the net at A-B, C-D. Difficult. (Continued on next page)

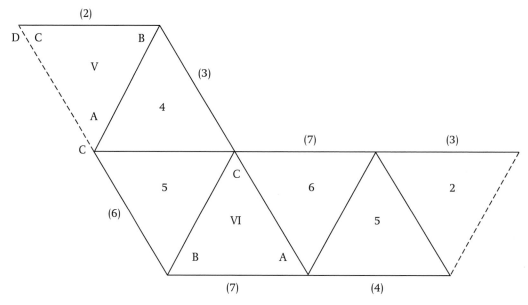

Fig. 4.16   (*Continued*)

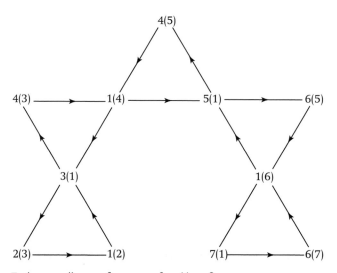

Fig. 4.17   Tuckerman diagram for a seven faced hexaflexagon.

4.5.2   A puzzle hexahexaflexagon

Fig. 4.19 shows the net for a puzzle hexahexaflexagon. The net is the same as the net shown in Fig. 4.12 except that the numbers on faces 1, 2 and 3 are replaced by a line on each of the triangles, and faces 4, 5, and 6 are blank. The challenge is, starting from the as assembled main position, to find the face where the lines form a rectangle. The solution can of course be found by using a Tuckerman traverse. Random flexing

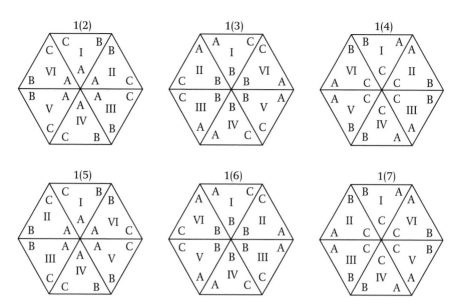

Fig. 4.18   Arrangements of face 1 on a seven faced hexaflexagon.

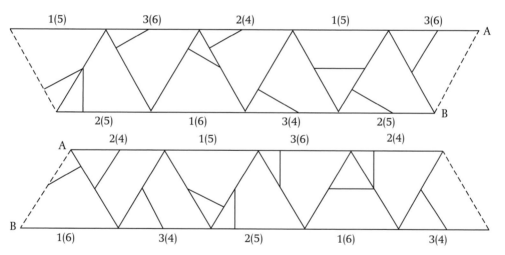

Fig. 4.19   Net for puzzle hexahexaflexagon with lines instead of face numbers and the reverse left blank. Join the two parts of the net at A-B. Face numbers are shown adjacent to triangles as an aid to assembly. For easy assembly write the face numbers lightly in pencil on the triangles, and erase after assembly.

is puzzling since some face arrangements appear in more than one main position.

The puzzle hexaflexagon was designed by first drawing a rectangle on face 1 of the seven faced hexaflexagon, with the hexaflexagon in main position 1(2). The patterns appearing in the six main positions in which face 1 appears are shown in Fig. 4.20. The main position codes are shown above the sketches. There are four distinct patterns, one of which appears

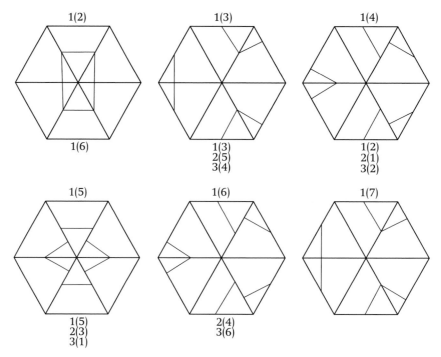

Fig. 4.20 Face appearances for the puzzle hexaflexagon. Main position codes above the sketches are for the seven faced hexaflexagon used during its design. Main position codes below the sketches are for the puzzle hexaflexagon itself.

twice in main positions 1(3) and 1(7), and another in main positions 1(4) and 1(6). The symmetry of these patterns means that they do not distinguish between different face arrangements. Next, the rectangle was drawn on face 1 of the puzzle hexahexaflexagon in main position 1(6). This ensures that the rectangle does appear. Then the pattern appearing in main position 1(3) was drawn on face 2 of the puzzle hexahexaflexagon in main position 2(5), and on face 3 in main position 3(4). Main position codes for the puzzle hexahexaflexagon are shown below the sketches. These show that one pattern appears five times, two patterns appear three times and the rectangle appears once.

In the seven faced hexaflexagon main positions including face 1 may be traversed (Fig. 4.17) in the sequence 1(2), 1(3), 1(4), 1(5), 1(6) and 1(7) or in its reverse. In the puzzle hexahexaflexagon main positions for face 1 may be traversed in the sequence 1(6), 1(3), 1(2) and 1(5) or in its reverse (Fig. 4.13). Similarly for face 2 the sequence is 2(5), 2(1), 2(3) and 2(4), and for face 3 it is 3(4), 3(2), 3(1) and 3(6). In each case the sequence of patterns is part of the sequence of patterns for the seven faced hexaflexagon. The design method used ensures that the part sequences for faces 2 and 3 do not include the rectangle.

5   Hexaflexagon variations

Extensive analysis has resulted in variations on the theme of hexaflexagons. Three are described in this chapter. These are: a different variety of flexagon, triangle flexagons; a different way of flexing hexaflexagons, the 'V-flex'; and origami like recreations with hexaflexagons.

A triangle flexagon is a variant of a hexaflexagon in which the number of sectors is reduced from three to two, correspondingly the number of leaves visible on a face of a main position is reduced from six to four. The sum of the leaf vertex angles at the centres of main positions is less than $360°$ and this results in slant main positions. These have the appearance of a square pyramid without a base. The dynamic behaviour of a triangle flexagon is similar to that of the corresponding hexaflexagon except that it isn't possible to traverse a complete cycle without disconnecting a hinge, refolding the flexagon and reconnecting the hinge.

The V-flex is a complicated flex which exploits the large number of degrees of freedom of all but the simplest types of hexaflexagon. Using the V-flex results in the faces of a hexaflexagon becoming mixed up. How to operate the V-flex is described in detail. By using both the V-flex and the pinch flex the hexahexaflexagon can be flexed to display numerous different main position faces. Individual leaf faces have to be identified in order to characterise these main position faces. Whether or not the V-flex is regarded as legitimate is a matter of personal taste. In this book it is not regarded as legitimate.

Flexagons are sometimes regarded as a branch of origami, the Japanese art of paper folding. Hexaflexagons can be folded into a variety of origami like shapes, and some examples are given.

5.1   Triangle flexagons

A triangle flexagon is a variant of a hexaflexagon in which the number of sectors is reduced from three to two and the number of pats in a main position from six to four. The net for a triangle flexagon is two thirds of the net for the corresponding hexaflexagon. In the triangle flexagon variant of a hexaflexagon there are four equilateral triangles about the centre of a main position. The sum of the vertex angles at the centre of a main position is therefore $240°$. Main positions are slant (Section 3.2)

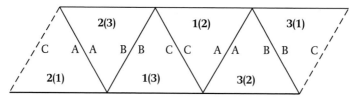

Fig. 5.1   Net for the single cycle triangle flexagon. Fold triangles numbered 3 together, then triangles numbered 1, and join the ends.

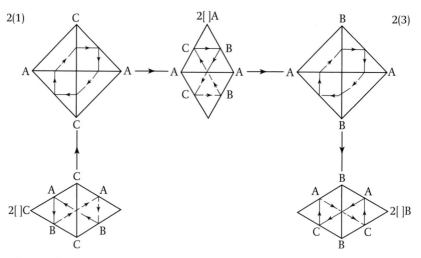

Fig. 5.2   Full map of the single cycle triangle flexagon.

and have the appearance of a square pyramid without a base as shown in Fig. 3.4. An intermediate position has the appearance of two triangles with a common side. Triangle flexagons are twisted bands but not Möbius bands. They are traversed using the pinch flex.

5.1.1   The single cycle triangle flexagon

The single cycle triangle flexagon corresponds to the trihexaflexagon (Section 4.2) so its net (Fig. 5.1) is two thirds of that for the trihexaflexagon (Fig. 4.3) and it is the only single cycle triangle flexagon. The full map is shown in Fig. 5.2. Main positions are shown looking down on the outsides of pyramids. Intermediate positions are identified by adding hinge letters to the codes. A hinge letter is the letter for the hinges running across the centre of an intermediate position. The flexagon figure is the same as that for the trihexaflexagon (Fig. 3.15).

As assembled the same face number (2) appears on the outside of the main and intermediate positions (Fig. 5.2). The dynamic properties of the single cycle triangle flexagon differ from those of the trihexaflexagon (Section 4.2) in that it isn't possible to traverse its cycle completely. All three of the expected intermediate positions can be reached, but only two of the expected main positions. The only way to reach the third main position is to disconnect a hinge, refold the flexagon so that either face number 1 or face number 3 appears on the outside of the pyramid, and reconnect the hinge. If this is done then a differently numbered and lettered full map is obtained.

At intermediate position 2[ ]A there are three degrees of freedom. One degree of freedom is folding about the central hinges, A-A. The other two degrees of freedom are opening the flexagon about hinges B-B and about C-C to reach a main position. At the other two intermediate positions there are only two degrees of freedom since it is only possible to reach one main position. Away from an intermediate position there is only one degree of freedom. There are no changes in the number of degrees of freedom to indicate that a main position has been reached, so a main position has to be defined as midway between the two adjacent intermediate positions.

5.2   The V-flex

The 'V-flex' (McLean 1979) is a complicated flex which exploits the large number of degrees of freedom of all but the simplest types of hexaflexagon. Using the V-flex results in faces becoming mixed up with different leaf numbers appearing on the same face. Whether or not the V-flex is regarded as legitimate is a matter of personal taste. In this book it is not regarded as legitimate. The reasons for this are that the V-flex is complicated, difficult to carry out, and not aesthetically satisfying. A lot of practice is needed to carry it out smoothly and without errors. By using both the V-flex and the pinch flex the hexahexaflexagon (Subsection 4.3.4) can be flexed to display 3940 different main position faces. Individual leaf faces have to be identified in order to characterise these main position faces.

5.2.1   V-flexing the hexahexaflexagon

To V-flex a paper model of a hexahexaflexagon, whose net is shown in Fig. 4.12, proceed as follows.

(a) Orient the hexahexaflexagon so that face 1 is uppermost, and also so that there are a pat consisting of a folded pile of four leaves at the top and a pat consisting of two leaves at the bottom, as shown by the adjacent numbers in Fig. 5.3(a).

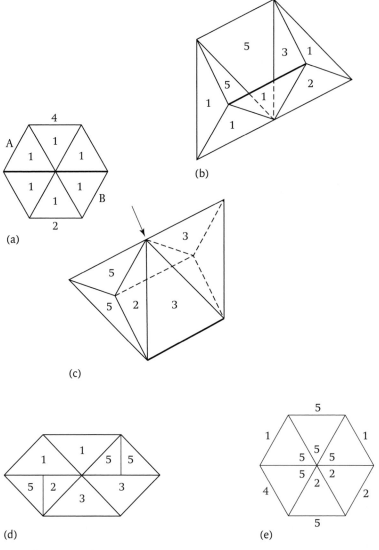

Fig. 5.3   The V-flex. (a) Initial main position. (b) Three pyramids. (c) Three pyramids inverted. (d) Top view of three pyramids inverted. (e) Final main position.

(b) Fold the top half of the hexahexaflexagon backwards along the horizontal hinges (heavy line in Fig. 5.3(a)) until it is at right angles to the lower half. Pull open the pats labelled A and B at the right angle fold to reach the three pyramids position shown in Fig. 5.3(b). The centre pyramid is a square pyramid, and the end pyramids are triangular pyramids. The insides of the centre and right hand pyramids are visible. The numbers on individual leaves in the figure are the hexahexaflexagon face numbers.

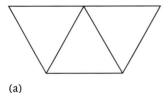

(a)

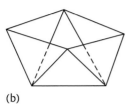

(b)

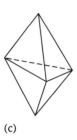

(c)

Fig. 5.4 Origami with the four faced hexaflexagon. (a) Folded in two. (b) Boat. (c) Triangular dipyramid.

(c) Rotate the front pat of the square pyramid about its upper side (heavy line) as far as it will go. This results in the inverted three pyramids position shown in Fig. 5.3(c). The inside of the left hand pyramid is visible. The rotated pat now has face number 3 visible, and the side about which it was rotated is now at the bottom (heavy line). A further view of the inverted three pyramids position, in the direction of the arrow, is shown in Fig. 5.3(d). In this view the insides of both end pyramids are visible.

(d) Complete the V-flex by collapsing the centre pyramid of the inverted three pyramids position (only one way works) to reach a new main position. Fig. 5.3(e) shows one of the new main position faces. On this face four leaves numbered 5 and two leaves numbered 2 are visible. On the other face four leaves numbered 1 and two leaves numbered 3 are visible. In the figure the adjacent numbers indicate the number of leaves in each pat. The threefold rotational symmetry of the pat structure of main positions, associated with pinch flexing the hexahexaflexagon (Subsection 4.3.4, Figs. 4.10 and 4.11), has been lost.

5.3   Origami with hexaflexagons

Flexagons are sometimes regarded as a branch of origami (Kenneway 1987). The pinch flex moves are rather like origami moves, and the V-flex moves needed to reach the three pyramids position shown in Fig. 5.3 are more so. Conversely, the large number of degrees of freedom of hexaflexagons means that origami like recreations are possible. As an example, start with the four faced hexaflexagon, whose net is shown in Fig. 4.7, in main position 1(2). Fold it in two (Fig. 5.4(a)). It can then be pulled apart at the top to form a boat (Fig. 5.4(b)). The sides of the boat have to be held together while doing this. In this position there are three degrees of freedom so it can be folded to several further configurations including the triangular dipyramid shown in Fig. 5.4(c).

Madachy (1968) gives detailed instructions on how to fold the hexahexaflexagon (Subsection 4.3.4) into various shapes. These include a square cup with a handle, twin cups, a pentagonal cup, twin triangular cups with a handle, a pair of corner-to-corner cups, a pair of side-by-side cups, a group of three pyramids (not the same as in Fig. 5.3), and a flower basket with a handle. A hexahexaflexagon can also be folded into an octahedron with a pair of opposite faces missing.

6   Square flexagons

Square flexagons were the second variety of flexagon to be discovered. They are less well understood than hexaflexagons, partly because their dynamic behaviour is more complex. In particular relatively little is known about numbers of distinct types with a given number of faces. The leaves of a square flexagon are squares. In appearance a main position of a square flexagon is flat and consists of four leaves each with a vertex at the centre so there are four pats and two sectors. The outline is a square. Some square flexagons are twisted bands and hence exist as enantiomorphic (mirror image) pairs. Enantiomorphs are not usually regarded as distinct types. The handedness of a square flexagon is only mentioned when this is needed for clarity. Some square flexagons are untwisted bands so there is only one form.

Hexaflexagons are relatively simple because there are only one type of cycle and only one type of link between cycles. By contrast square flexagons have three different types of cycle and two types of link between cycles. There are three distinct types of single cycle square flexagon. Two of these have four faces and will traverse a complete 4-cycle. The third has three faces, and is incomplete in that it will only traverse an incomplete 3-cycle.

A 'main position link' between two cycles in a multicycle square flexagon is analogous to the type of link which occurs in multicycle hexaflexagons. Some statistics on numbers of distinct types of square flexagon with main position links are given. In some square flexagon intermediate positions it is possible to open an intermediate position into a 'box position'. It is sometimes possible to link square flexagons at box positions to create a 'box position link' between cycles. A box flex is then used to traverse between the linked cycles. The complex dynamic behaviour of square flexagons facilitates the construction of puzzle flexagons. A puzzle square flexagon with one main position link and a pair of box position links is described. Another puzzle square flexagon uses the 'interleaf flex' which is a variant of the box flex.

The design of nets for specific types of square flexagons whose dynamic behaviour is known is more difficult than for hexaflexagons. Starting from a known design, practical methods based on paper models make it possible to add or delete cycles and faces, and hence design square flexagons with desired dynamic behaviours. These methods are described in detail.

There are eight possible arrangements of the leaves which make up a face of a square flexagon, but if only main position links are present there are only four. These are illustrated by a specially designed five faced square flexagon which includes all four of these arrangements of one of its faces.

6.1   Single cycle square flexagons

There are three distinct types of single cycle square flexagon. Two of these have four faces and will traverse a complete 4-cycle. The third has three faces, and is incomplete in that it will only traverse an incomplete 3-cycle.

6.1.1   The irregular single cycle square flexagon revisited

The irregular single cycle square flexagon (Subsection 3.4.1) has four faces and will traverse a complete 4-cycle. Its net is shown in Fig. 3.9. It is called irregular because, as can be seen on the full map (Fig. 3.8), main positions have two distinct types of pat structure. Also, the inscribed part of the flexagon figure (Fig. 3.14) is a crossed quadrilateral, which is an irregular polygon. This particular type of square flexagon is an untwisted band. This is easily seen by bending the leaves in either main position 2(1) or main position 4(3). More formally in, say, main position 2(1) the top left and bottom right pats have a right handed twist of $-180°$ and the other two pats have a left handed twist of $180°$. The total twist around the flexagon is therefore zero, hence it is an untwisted band.

An irregular single cycle flexagon may be defined as a single cycle flexagon whose flexagon figure is an irregular polygon with the same number of sides as the circumscribing polygon. There is a tendency to regard irregular single cycle flexagons as somehow not quite respectable. Conrad and Hartline (1962) call them 'improper' flexagons.

6.1.2   The regular single cycle square flexagon

Fig. 6.1 shows the net for the second type of single cycle square flexagon with four faces. It will traverse a complete 4-cycle. It is a twisted band but not a Möbius band. The full map (Fig. 6.2) shows that the sequence of faces visible in main and intermediate positions is identical with that for the irregular single cycle square flexagon (Fig. 3.8). However, this square flexagon is regular in that all four main positions have the same pat structure. In the main positions pats are alternate single leaves and folded piles of three leaves. The pats consisting of piles of three leaves have a left handed twist of $-360°$ so this is the left handed enantiomorph. The flexagon figure (Fig. 6.3) is a square, which is a regular polygon, with

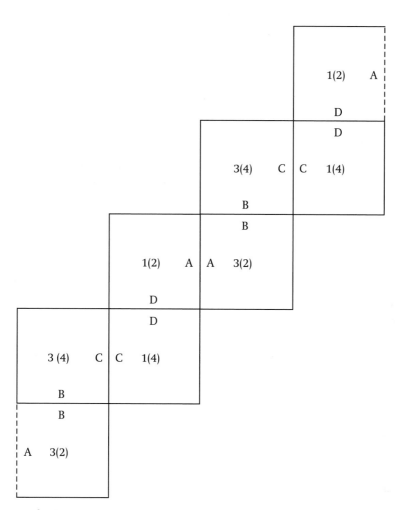

Fig. 6.1 Net for the regular single cycle square flexagon.

vertices on the midpoints of a circumscribing square. The simplified map, Tuckerman diagram, and the intermediate position map are the same as those for the irregular single cycle square flexagon (Figs. 3.11, 3.12 and 3.13). Information that differentiates between the two types of square flexagon has been lost in the simplification.

A regular single cycle flexagon may be defined as a single cycle flexagon whose flexagon figure is a regular polygon with the same number of sides as the circumscribing polygon.

6.1.3 The incomplete single cycle square flexagon revisited

The simple square flexagon described in Section 1.3 is actually the incomplete single cycle square flexagon. It is the third type of single cycle

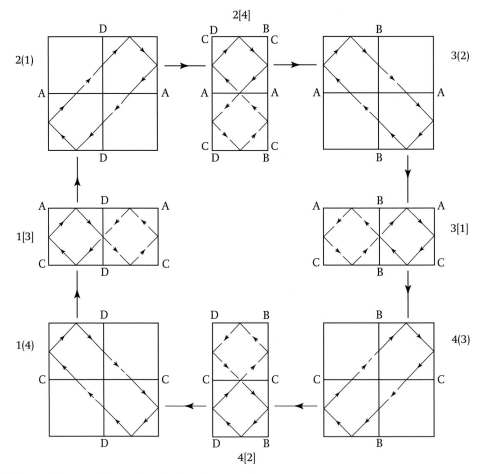

Fig. 6.2   Full map of the regular single cycle square flexagon.

Fig. 6.3   Flexagon figure for the regular single cycle square flexagon.

square flexagon, and has three faces and three main positions. The net shown in Fig. 6.4 is the same as that shown in Fig. 1.3 except that hinge letters have been added. As shown in the full map (Fig. 6.5) the incomplete single cycle square flexagon can only traverse part of a cycle: an incomplete 3-cycle. The two main positions are identical although this is not immediately obvious from the figure. The identity becomes clear if one of the main positions is turned over and its circuit redrawn. In the code for intermediate position 2[ ] the empty square brackets show

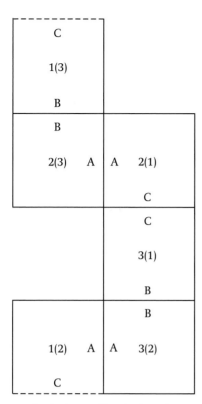

Fig. 6.4   Net for the incomplete single cycle square flexagon.

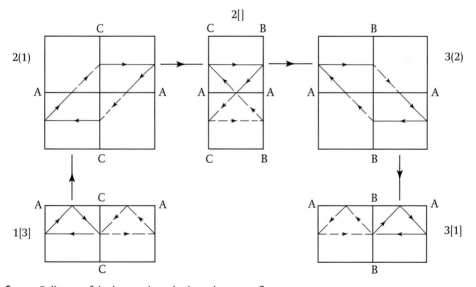

Fig. 6.5   Full map of the incomplete single cycle square flexagon.

Fig. 6.6   Flexagon figure for the incomplete single cycle square flexagon.

(a)   (b)

Fig. 6.7   Flexagon figure for the regular single cycle square flexagon. (a) Arbitrary direction selected. (b) Disconnected at an arbitrary vertex.

that it is not possible to open the square flexagon into a box position (Fig. 3.10). It is possible to do this in intermediate positions 1[3] and 3[1]. The flexagon figure (Fig. 6.6) is a right angle triangle with vertices on three of the midpoints of a circumscribing square.

An incomplete single cycle flexagon may be defined as a single cycle flexagon whose flexagon figure is a polygon with fewer sides than the circumscribing polygon.

6.1.4   Degrees of freedom

A main position of a square flexagon has two degrees of freedom. For example main position 2(1) in Fig. 6.2 can be rotated about either hinges A-A or hinges D-D. Once in motion between a main position and an intermediate position there is one degree of freedom. An intermediate position has a maximum of four degrees of freedom. For example intermediate position 2[4] in Fig. 6.2 can be rotated about hinges A-A, opened about hinges B-B or D-D, or opened up into an open ended box.

6.2   Reconstruction from a flexagon figure

A single cycle square flexagon, defined in terms of its flexagon figure, may be reconstructed using the following procedure. This reverses the procedure used to define a flexagon figure (Section 3.5). The regular single cycle square flexagon is used as an example. Its flexagon figure is shown in Fig. 6.3.

Choose an arbitrary direction around the flexagon figure, as in Fig. 6.7(a), and disconnect it at an arbitrary vertex, as at the bottom of Fig. 6.7(b). The circumscribing polygon is a square. This shows that the leaves of the flexagon are squares. The inscribed polygon has four sides so a pat consisting of a folded pile of four paper leaves is required. Fig. 6.7(b) indicates how the individual leaves are to be connected. Starting at the bottom of the pile join the first and second leaves with transparent adhesive tape at the left side, the second and third leaves at the top, and the third and fourth leaves at the right side. Make a second identical pat, turn it over and place it against the first pat so that the sides corresponding to the disconnected vertex are adjacent. Join the top leaves with transparent adhesive tape, then the bottom leaves. This configuration is an intermediate position of the desired square flexagon.

If the arbitrary direction, clockwise in Fig. 6.7(a), is reversed so that it becomes anticlockwise then the enantiomorph is obtained. If clockwise and anticlockwise directions do not lead to distinct results, as in Fig. 3.14, then the square flexagon is an untwisted band and there is no enantiomorph. With due attention to detail this procedure may also be used for single cycle flexagons whose leaves are other types of polygon.

The procedure can lead to only one distinct type of flexagon from a given flexagon figure. Figs. 3.14, 6.3 and 6.6 show the only possible distinct ways in which a polygon can be inscribed in a square, with vertices on the midpoints of the sides of the square. These are therefore the only possible flexagon figures, so there can be no other types of single cycle square flexagon. Similarly, Fig. 3.15 shows the only way in which a polygon can be inscribed in an equilateral triangle with vertices on the midpoints of the sides of the triangle. Hence the trihexaflexagon is the only possible single cycle hexaflexagon. Conrad and Hartline (1962) use a similar argument and reach the same conclusions for single cycle flexagons that can traverse a complete cycle.

6.3   Square flexagons with main position links

A main position link between two cycles in a multicycle square flexagon is analogous to the type of link which occurs in multicycle hexaflexagons. Hence paper models of multicycle square flexagons containing main position links may, with due attention to detail, be constructed using the synthetic method (Subsection 4.4.2).

6.3.1   Main position links

To link paper models of two square flexagons using a main position link the procedure is as follows. Flex each square flexagon to the main position which is to be linked. This must be one in which alternate pats are single leaves. A regular single cycle square flexagon can be linked at any of its four main positions. However, an irregular single cycle square flexagon can only be linked at two main positions because there are only two which have alternate single leaves, for example main positions 3(2) and 1(4) in Fig. 3.8. An incomplete single cycle square flexagon can be linked at either of its main positions. Cut the flexagons along the hinge lines between leaves, and remove the single leaves. Keep the remaining pats in their original relationship to one another. Mark the cut sides of leaves in the pats for future reference. Assemble the common main position of the square flexagons being linked by placing the remaining pats from each of the original square flexagons alternately. Join the marked side of the top leaf of each pat to the marked side of the bottom leaf of an adjacent pat. If this is not possible then change the handedness of one of the square flexagons being linked. If it is a regular single cycle square flexagon, or an incomplete single cycle square flexagon, this may be done by using the enantiomorph. If it is an irregular single cycle square flexagon then the handedness of the pats may be changed by using the other available main position. This works because the two available main positions are enantiomorphs. Renumbering of faces will usually be needed. In linking the two square flexagons four leaves and hence two faces are removed.

6.3.2   The square flexagon with two regular cycles and a main position link

As an example of a main position link, Fig. 6.8 shows the net for the two cycle square flexagon formed by linking two left handed regular single cycle square flexagons (Subsection 6.1.2). The common main position assembled during linking is shown in Fig. 6.9. It is main position 2(1) in the simplified map (Fig. 6.10). The resulting flexagon is also a left handed twisted band. The internal structure of the pats is more complex than for the corresponding two cycle hexaflexagon (Subsection 4.3.2).

From the common main position, 2(1) in Fig. 6.10, the square flexagon with two regular cycles, main position link, can be flexed to four different main positions. Once away from the common main position the dynamic behaviour of each cycle has to be considered separately. Consider cycle A at, say, intermediate position 2[4]. Fig. 6.11 shows details of the pat structure in this intermediate position. The leaves represented by broken lines in the upper square of the figure and by solid lines in the lower square belong to cycle B. Within cycle A, away from main position 2(1), these cycle B leaves are locked together by reverse folds, and behave as if

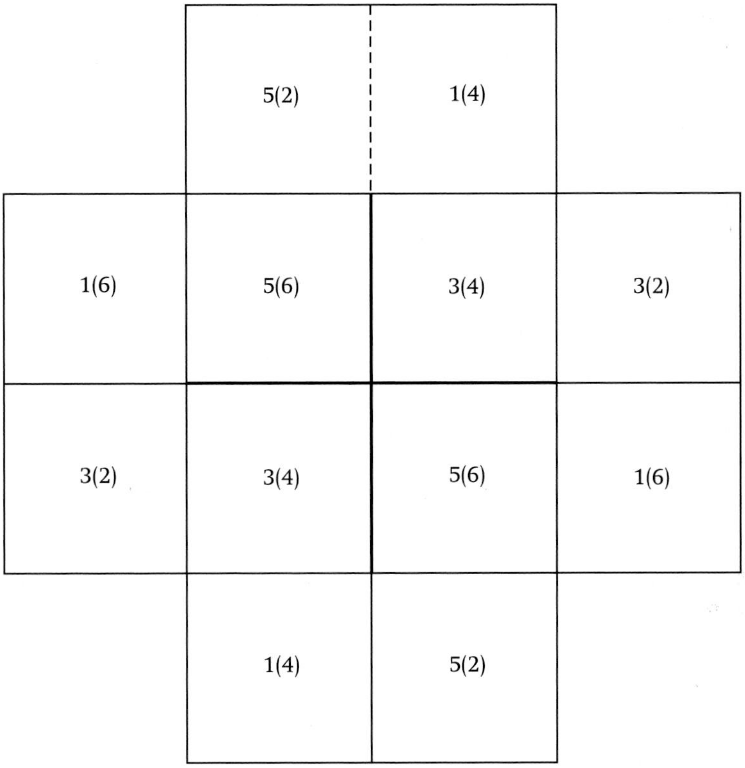

5(2)	1(4)

1(6)	5(6)	3(4)	3(2)

3(2)	3(4)	5(6)	1(6)

1(4)	5(2)

Fig. 6.8   Net for the square flexagon with two regular cycles, main position link. Cut along heavy and dashed lines.

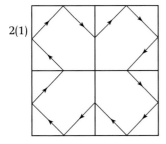

2(1)

Fig. 6.9   Diagram of main position 2(1), square flexagon with two regular cycles, main position link.

they were a single leaf. In effect, intermediate position 2[4] is identical to intermediate position 2[4] in Fig. 6.2. The dynamic behaviour of cycle A is therefore identical to that of the regular single cycle square flexagon (Fig. 6.1). Cycle A may therefore be described as regular. As the two cycles are identical cycle B is also regular.

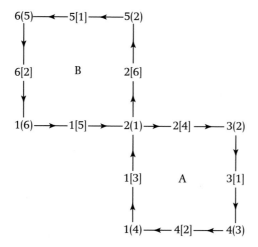

Fig. 6.10   Simplified map of the square flexagon with two regular cycles, main position link.

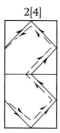

Fig. 6.11   Diagram of intermediate position 2[4], square flexagon with two regular cycles, main position link.

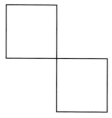

Fig. 6.12   Flexagon diagram for the square flexagon with two regular cycles, main position link.

6.3.3   Square flexagons with regular cycles and main position links

Square flexagons with regular cycles and main position links are the square flexagon equivalent of hexaflexagons. Their dynamic behaviour can be conveniently represented by flexagon diagrams. The number of faces is two plus twice the number of squares in the flexagon diagram. Fig. 6.12 shows the flexagon diagram for the square flexagon with two regular cycles and a main position link. A Tuckerman traverse (Subsection 4.3.3) guarantees that all possible main positions are visited.

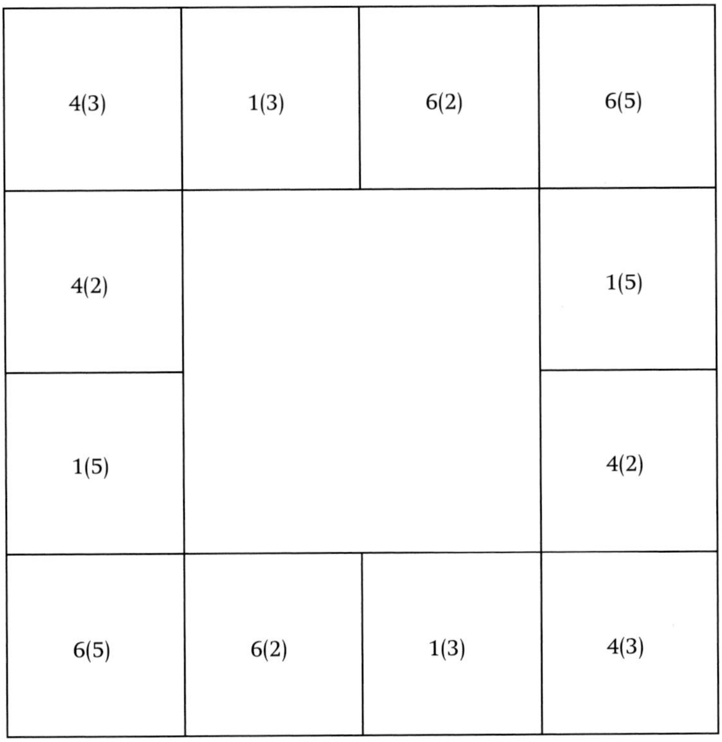

4(3)	1(3)	6(2)	6(5)
4(2)			1(5)
1(5)			4(2)
6(5)	6(2)	1(3)	4(3)

Fig. 6.13 Net for the square flexagon with two irregular cycles, main position link. This is an untwisted band and can be assembled, without cutting, by bending the leaves.

6.3.4 The square flexagon with two irregular cycles and a main position link

Apart from the square flexagon with two regular cycles and a main position link (Subsection 6.3.2) there are two other types of square flexagon with two complete cycles and a main position link. Fig. 6.13 shows the net for the square flexagon with two irregular cycles and a main position link. This flexagon is an untwisted band and the net is a ring. It can be assembled, without cutting, by bending the leaves. There are other square flexagons for which this is possible (Neale 1999).

6.3.5 The square flexagon with one regular cycle and one irregular cycle and a main position link

Fig. 6.14 shows the net for the square flexagon with one regular cycle, one irregular cycle and a main position link. The net is a hybrid between those shown in Figs. 6.8 and 6.13. The dynamic behaviour of all three of the square flexagons whose nets are shown in Figs. 6.8, 6.13 and 6.14 is in general the same, but differs in detail. The simplified map (Fig. 6.10) is the same for all three. To distinguish between them the simplified

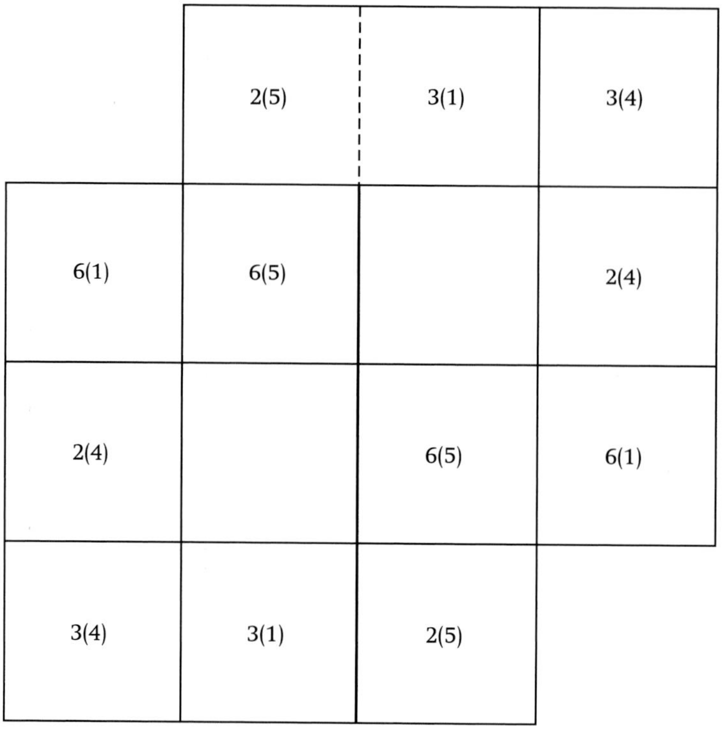

	2(5)	3(1)	3(4)
6(1)	6(5)		2(4)
2(4)		6(5)	6(1)
3(4)	3(1)	2(5)	

Fig. 6.14   Net for the square flexagon with one regular and one irregular cycle, main position link. Cut along the dashed line and cut out the unnumbered squares.

map would have to be annotated to show the types of cycle present. The flexagon diagram (Fig. 6.12) is also the same for all three square flexagons.

6.3.6   Deleting faces from square flexagons

As with hexaflexagons (Subsection 4.4.1) a straightforward practical method of deriving the net for a flexagon with desired dynamic properties is to delete faces from a paper model of an appropriate known net. This is particularly convenient for square flexagons with main position links since the result can often be visualised without actually using a paper model.

Fig. 6.4 shows the net for the left handed enantiomorph of the incomplete single cycle square flexagon, which has three faces. It can be derived from the net for the left handed enantiomorph of the regular single cycle square flexagon (Fig. 6.1) by gluing together the two pairs of faces of leaves numbered 4. Alternatively, fasten the leaves together temporarily with paper clips and copy the net. Gluing together any of the other numbers on the faces of leaves also results in the left handed enantiomorph.

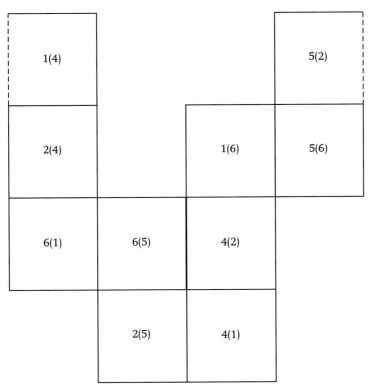

Fig. 6.15   Net for the square flexagon with one regular and one incomplete cycle, main position link. Cut along heavy line.

If faces numbered 2 are to be glued together then the net first needs to be cut at a hinge B or C and the disconnected hinge A joined. A net for the incomplete single cycle square flexagon can also be derived from the net for the irregular single cycle square flexagon (Fig. 3.9). Whether the right handed or left handed enantiomorph is generated depends on which face number is selected. Gluing together faces of leaves numbered 1 or 4 results in the left handed enantiomorph, and gluing together faces numbered 2 or 3 results in the right handed enantiomorph.

6.3.7   The square flexagon with one regular cycle and one incomplete cycle and a main position link

Deleting face 3 from the net for the square flexagon with two regular cycles and a main position link (Fig. 6.8) results in the square flexagon with one regular cycle and one incomplete cycle and a main position link. The net and simplified map are shown in Figs. 6.15 and 6.16. Cycle A (Figure 6.10) has become an incomplete cycle. It should be noted that in the presence of incomplete cycles a Tuckerman traverse (Subsection

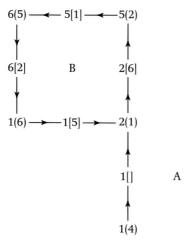

Fig. 6.16   Simplified map of the square flexagon with one regular and one incomplete cycle, main position link.

4.3.3) cannot be used to guarantee that all possible main positions are visited.

6.3.8   The square flexagon with one irregular cycle and one incomplete cycle and a main position link

Deleting face 5 from the net for the square flexagon with two irregular cycles and main position link (Fig. 6.13) results in the square flexagon with one irregular cycle and one incomplete cycle and a main position link. The net is shown in Fig. 6.17. The simplified map (Fig. 6.18) is similar to that shown in Fig. 6.16 but the face numbers and cycle letters are different.

6.3.9   Square flexagons with two incomplete cycles and a main position link

Two different types of square flexagons with two incomplete cycles and a main position link may be derived from the net for the square flexagon with one regular and one incomplete cycle and a main position link (Fig. 6.15). Deleting face 5 leads to the net shown in Fig. 6.19 and deleting face 6 to Fig. 6.20. The corresponding simplified maps are shown in Figs. 6.21 and 6.22. In Fig. 6.21 the simplified map doesn't differentiate between the two different intermediate positions. A third type may be derived from the net for the square flexagon with two irregular cycles and a main position link (Fig. 6.13) by deleting faces 3 and 5. The simplified map is the same as that shown in Fig. 6.21 but the pat structure is different.

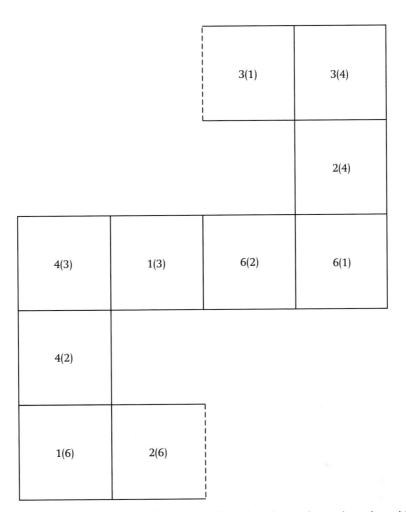

Fig. 6.17 Net for the square flexagon with one irregular and one incomplete cycle, main position link.

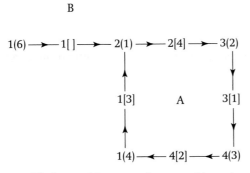

Fig. 6.18 Simplified map of the square flexagon with one irregular and one incomplete cycle, main position link.

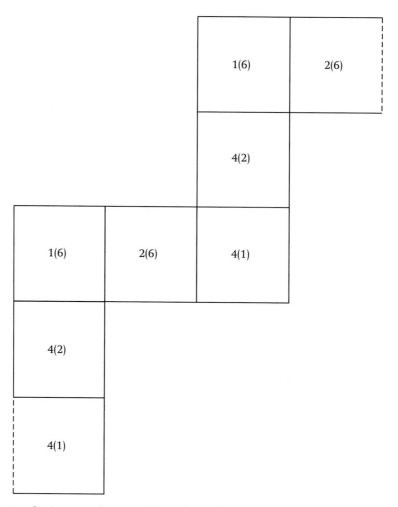

Fig. 6.19   Net for the square flexagon with two incomplete cycles, main position link, face 5 deleted from Fig. 6.15.

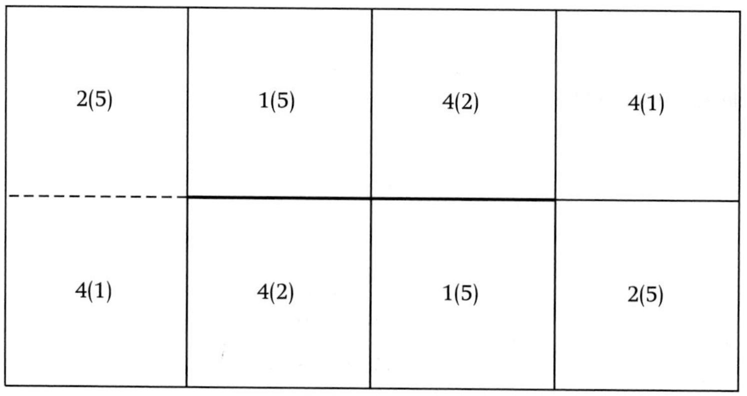

2(5)	1(5)	4(2)	4(1)
4(1)	4(2)	1(5)	2(5)

Fig. 6.20   Net for square flexagon with two incomplete cycles, main position link, face 6 deleted from Fig. 6.15. Cut along heavy and dashed lines.

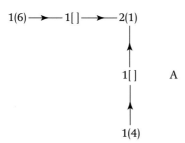

Fig. 6.21   Simplified map of square flexagon with two incomplete cycles, main position link, face 5 deleted from Fig. 6.15.

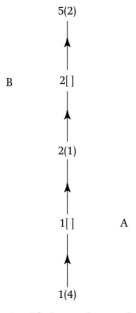

Fig. 6.22   Simplified map of square flexagon with two incomplete cycles, main position link, face 6 deleted from Fig. 6.15.

6.4   Distinct types of square flexagon with main position links

Table 6.1 shows the number of distinct types (enantiomorphs are not regarded as distinct) of square flexagons with regular cycles only and main position links. These numbers were obtained by considering possible flexagon diagrams. Only even numbers of faces are possible. The table also shows that there is only 1 type of square flexagon with irregular cycles only and main position links for each number of faces. This is because of

Table 6.1 *Numbers of distinct types of square flexagon, complete cycles only, main position links.*

Number of faces	Number of cycles	Number of types Regular cycles only	Number of types Irregular cycles only
4	1	1	1
6	2	1	1
8	3	2	1
10	4	5	1
12	5	16	1
14	6	60	1

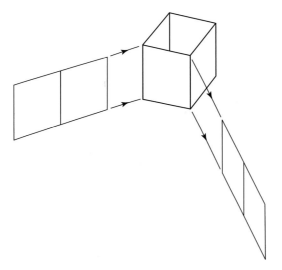

Fig. 6.23   The box flex.

the restricted number of main positions where main position links are possible.

If all types of cycle are considered then there are 2 distinct types of multicycle square flexagon with four faces, 6 distinct types with five faces, and 29 distinct types with six faces. The increase in numbers of distinct types with the number of faces is much more rapid than for hexaflexagons (Table 4.1).

6.5   Square flexagons with box position links

In some square flexagon intermediate positions it is possible to open an intermediate position into a box position (Fig. 3.10). In a 'box flex' (Fig. 6.23) an intermediate position is opened into an open ended box and then closed into another version of the intermediate position. For example, in intermediate position 2[4], Fig. 6.2, a box flex would interchange the positions of hinges A-A and C-C. It is sometimes possible to link square flexagons at box positions to create a box position link between cycles. A box flex is then used to traverse between the linked cycles. The box flex is sometimes called the 'tube flex', and a flexagon capable of being box flexed a 'tubulating flexagon'.

In a box position link two square flexagons are linked at two pairs of intermediate positions of the form a[b] and b[a], for example intermediate positions 1[3] and 3[1] in Figs. 6.2 and 6.5. A fundamental difference between main position links and box position links is that main position links occur singly whereas box position links always occur in pairs. Main position links and box position links are mutually exclusive in the sense that it may be possible to link two square flexagons by one or the other, but not by both.

6.5.1   Box position links

The synthetic method using paper models (Subsections 4.4.2 and 6.3.1) is the easiest way to construct box position links. There are some restrictions on what square flexagons can be linked. To link two square flexagons, flex each of the square flexagons to be linked to the box position which is to be linked. This must be one in which alternate pats are single leaves. These pats are marked 1 in Fig. 6.24, top. Cut the flexagons along the hinge lines between leaves, and remove the single leaves. Keep the remaining pats in their original relationship to one another (Fig. 6.24, centre). Mark the cut edges of leaves in the pats for future reference. Assemble the common box position of the new square flexagon by assembling the remaining pats from each square flexagon to form opposite faces of an open ended box. The arrangement must be as indicated by the letters on the pats in Fig. 6.24, bottom. Join the cut side of a leaf at the end of a pat inside the box to the cut side of a leaf in an adjacent pat at its end outside the box. It is usually easy to see what needs to be done to achieve this. If one of the square flexagons being linked is turned upside down then a different result may be obtained.

6.5.2   Square flexagons with two incomplete cycles and a pair of box position links

There are two distinct types of square flexagon with two incomplete cycles and a pair of box position links. The nets for these are shown in Figs. 6.25

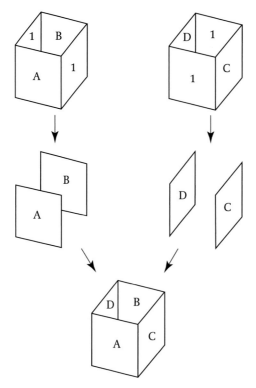

Fig. 6.24   Construction of box position links.

and 6.26. They are formed by linking two incomplete single cycle square flexagons in two different ways. The simplified map (Fig. 6.27) is the same for both types, but the pat structure is different. Small squares are used to indicate a box flex. By using two box flexes all four main positions can be visited in cyclic order. The two forms of an intermediate position before and after a box flex are distinguished on the simplified map by adding a letter denoting the central hinges to the code.

As the four main positions can be visited in cyclic order the flexagons could be described as another type of single cycle square flexagon. However, as the cycle is actually made up from two incomplete cycles it is best described as a 'pseudo-4-cycle'. The simplified map has been drawn to make this clear.

6.5.3   The square flexagon with two irregular cycles and a pair of box position links

The net for the square flexagon with two irregular cycles and a pair of box position links is shown in Fig. 6.28. The simplified map (Fig. 6.29) shows that the dynamic behaviour is quite complex even though the

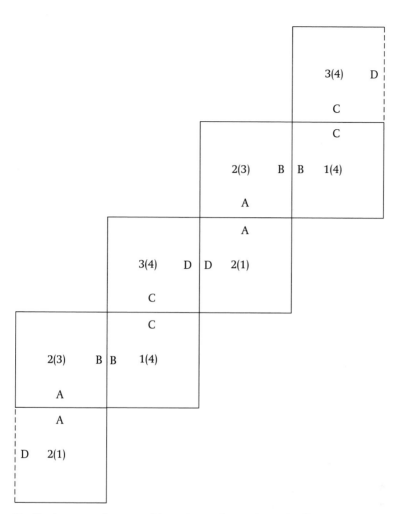

Fig. 6.25 Net for the square flexagon with two incomplete cycles, pair of box position links, first type.

D 2(1)	3(4) D	D 2(1)	3(4) D
C	A	C	A
C	A	C	A
2(3) B	B 1(4)	2(3) B	B 1(4)

Fig. 6.26 Net for the square flexagon with two incomplete cycles, pair of box position links, second type. Cut along heavy lines.

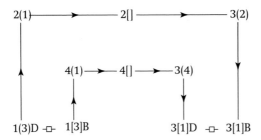

Fig. 6.27   Simplified map of square flexagons with two incomplete cycles, pair of box position links.

A 1(4)	6(3) A	A 1(4)	6(3) A
B	F	B	F
B	F	B	F
2(4)	6(5)	2(4)	6(5)
C	E	C	E
C	E	C	E
2(3) D	D 1(5)	2(3) D	D 1(5)

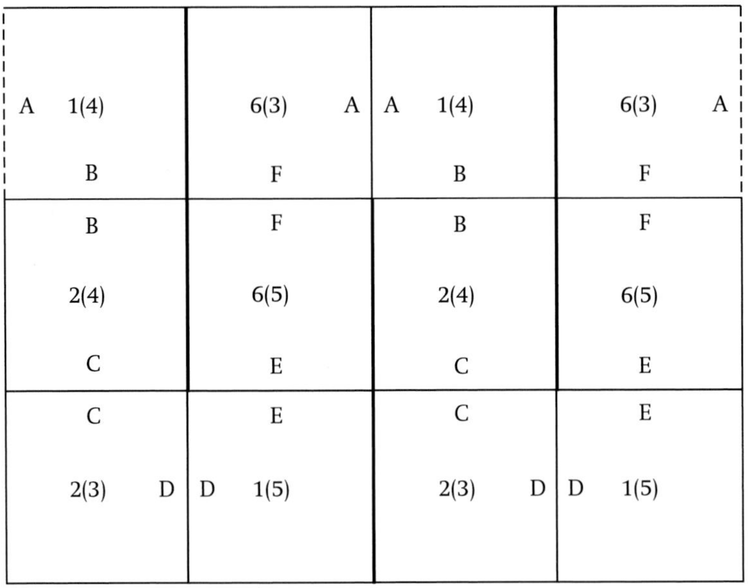

Fig. 6.28   Net for the square flexagon with two irregular cycles, pair of box position links. Cut along heavy lines. Difficult.

flexagon has only six faces. There are eight main positions and several different routes by which it can be traversed from, say, main position 5(1) to main position 6(3). The dynamic behaviour is more complex than that of the square flexagon with two irregular cycles and a main position link (Subsection 6.3.4).

6.5.4   A puzzle square flexagon

Figs. 6.30 and 6.31 show the net and simplified map for a square flexagon with one main position link and a pair of box position links. It is presented as a puzzle square flexagon by Mitchell (1999) with the challenge: starting from main position 4(3) find main position 5(6). It is an untwisted band and can be assembled by bending the leaves without

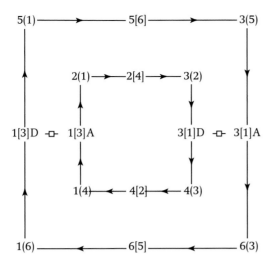

Fig. 6.29 Simplified map of the square flexagon with two irregular cycles, pair of box position links.

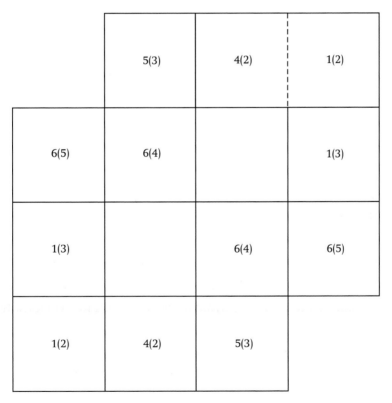

Fig. 6.30 Net for a square flexagon with one main position link and a pair of box position links. Cut along dashed line and remove unnumbered squares. By bending the leaves can be assembled without cutting along the dashed line. Difficult.

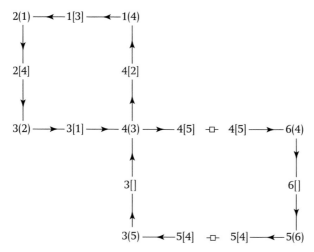

Fig. 6.31   Simplified map of a square flexagon with one main position link and a pair of box position links.

cutting along the dashed line. It consists of an irregular cycle (main positions 4(3), 1(4), 2(1) and 3(2)) with a main position link to an incomplete cycle (main positions 4(3) and 3(5)) which in turn has a pair of box position links to another incomplete cycle. This second incomplete cycle (main positions 6(4) and 5(6)) includes face 6. Starting from the irregular cycle it is quite difficult to find a box position leading to face 6. To emphasise this point hinge letters have been deliberately omitted from the net.

6.6   Square flexagon face arrangements

The arrangements of the squares making up a face of a square flexagon change as it is flexed from one main position to another. As with hexaflexagons (Section 4.5) these changes are used both as the basis of puzzles (Mitchell 1999) and as the basis of decorative schemes (Mitchell 1999).

6.6.1   A five faced square flexagon

If only main position links are present then there are four possible arrangements of the leaves which make up a face of a square flexagon. The simplest square flexagon on which all four possible arrangements of a face occur is the five faced square flexagon whose net is shown in Fig. 6.32. The Tuckerman diagram (Fig. 6.33) shows that face 1 is visible in all four main positions. Face 1 is not numbered. Instead letters are used to identify the vertices of leaves and Roman numerals to identify

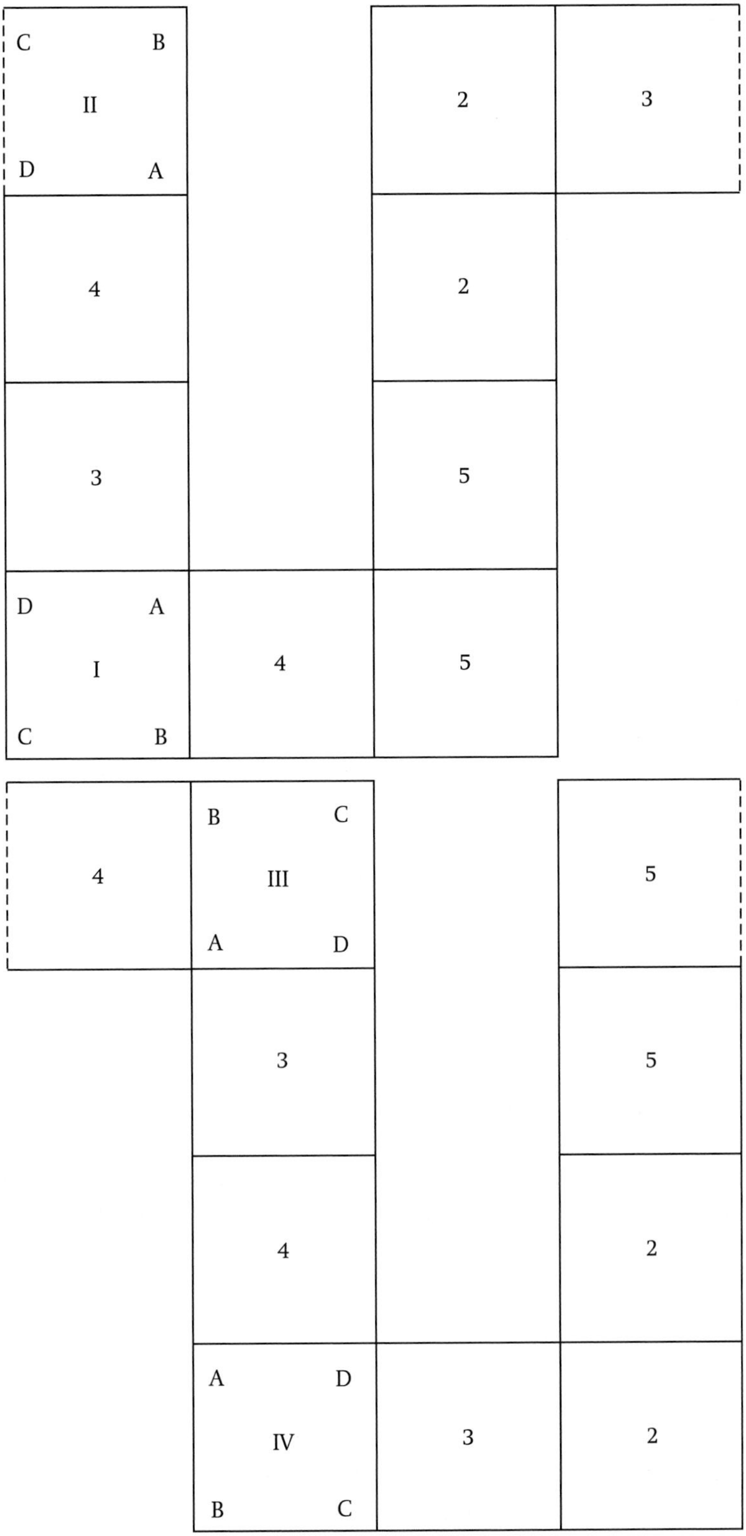

Fig. 6.32 Net for a five faced square flexagon. Both sides of the net are shown. Face 1 is not numbered. Fold leaves together in the order 5, 2, 3. Difficult.

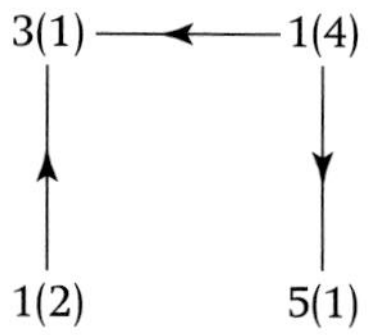

Fig. 6.33   Tuckerman diagram for a five faced square flexagon.

1(2)			1(3)			1(4)			1(5)		

1(2)

C	B	B	C
IV		I	
D	A	A	D
D	A	A	D
III		II	
C	B	B	C

1(3)

D	A	A	D
I		IV	
C	B	B	C
C	B	B	C
II		III	
D	A	A	D

1(4)

A	D	D	A
IV		I	
B	C	C	B
B	C	C	B
III		II	
A	D	D	A

1(5)

B	C	C	B
I		IV	
A	D	D	A
A	D	D	A
II		III	
B	C	C	B

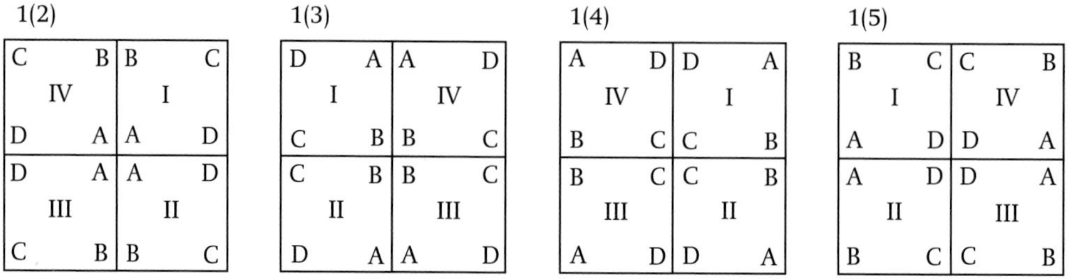

Fig. 6.34   Arrangements of face 1 on a five faced square flexagon.

individual leaves. The orientation of these letters and numerals is not regarded as having any significance. The changes in the arrangements of faces as the square flexagon is flexed from main position 1(2) to main position 1(5) are shown in Fig. 6.34. The letters at the centre of a face are in the sequence ABCD, and adjacent squares rotate in opposite directions from one main position to the next. The Roman numerals run alternately clockwise and anticlockwise. In combination these changes mean that there can only be four possible arrangements, all of which are shown in the figure.

If there are box position links then it is possible for either clockwise or anticlockwise Roman numerals to be associated with any face centre letter. There are then eight possible arrangements of the squares on a face. The simplest square flexagon in which all eight possible face arrangements occur has 10 faces. It is very difficult to handle so its net is not shown.

6.7   The interleaf flex

The 'interleaf flex' is a variant of the box flex (Section 6.5) which is possible with some types of square flexagon. It is described by Mitchell (1999) but not given a name. Using the interleaf flex results in some faces becoming mixed up, with two different leaf numbers appearing on the same face. Whether or not the interleaf flex is regarded as legitimate is a matter of personal taste. In this book it is not regarded as

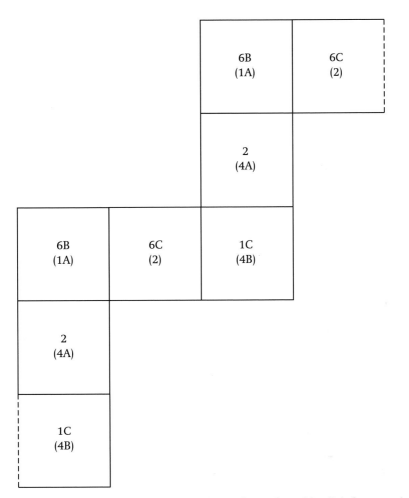

Fig. 6.35   Net for square flexagon with two incomplete cycles, main position link, faces 3 and 5 deleted from Fig. 6.13. Ignore face letters during assembly. Difficult.

legitimate. The reasons for this are that the interleaf flex is difficult to carry out, involves severe bending of the leaves and is not aesthetically satisfying.

6.7.1   Interleaf flexing a square flexagon with two incomplete cycles and a main position link

Fig. 6.35 shows the net for a square flexagon with two incomplete cycles and a main position link. It is the third type of square flexagon mentioned in Subsection 6.3.9 and was derived from the net shown in Fig. 6.13 by deleting faces 3 and 5. On the net leaves are identified by face letters as well as by face numbers. The simplified map is shown in Fig. 6.36. As

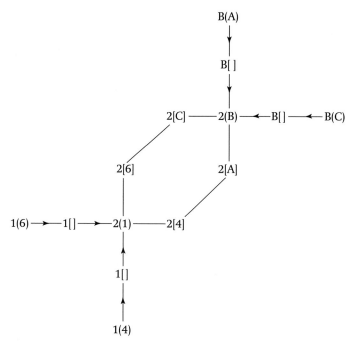

Fig. 6.36 Simplified map of square flexagon with two incomplete cycles, main position link, faces 3 and 5 deleted from Fig. 6.13.

assembled, ignoring face letters, the square flexagon is in main position 2(1); main positions 1(4) and 1(6) may be reaching using pinch flexes. Thus the bottom left part of the simplified map is identical to the simplified map shown in Fig. 6.21.

To operate the interleaf flex start from main position 2(1) and box flex to, say, box position 2[4]. In this position two of the leaves, on opposite sides of the inside of the box, are marked 4B. The other two leaves are marked 4A. Interleave pairs of leaves, one pair at a time, so that the leaves marked 4B are replaced by leaves marked 1A and box position 2[A] is reached. Complete the flex by box flexing to main position 2(B). On face B two of the leaves are marked 4B and the other two 6B. It is also possible to flex from main position 2(1) to main position 2(B) via box positions 2[6] and 2[C]. It is therefore possible to traverse a pseudocycle.

From main position 2(B) main positions B(A) and B(C) may be reached by using pinch flexes. The top right part of the simplified map (Fig 6.36) is identical, except for the use of face letters instead of numbers, to the simplified map shown in Fig. 6.21. In other words the interleaf flex transforms this particular square flexagon with two incomplete cycles and a main position link into another version of itself but with faces made up from different combinations of leaves. This square flexagon is presented as a puzzle by Mitchell (1999) with the challenge, starting from main position 2(1) find main position 2(B). He uses an ingenious decoration scheme and presentation to disguise the fact that faces become mixed up.

7 Introduction to convex polygon flexagons

A formal textbook on flexagons, after some preliminaries, would probably start with a chapter with this somewhat daunting title, or something similar. In this book the chapter has been deferred to this point because convex polygon flexagons are nothing more than generalisations of the square flexagons and triangle flexagons described earlier. Understanding of convex polygon flexagons in general is incomplete.

There is an infinite family of convex polygon flexagons. Varieties are named after the constituent polygons. A feature of some varieties of convex polygon flexagon is that there may be more than one type of main position and more than one type of complete cycle. It then becomes necessary to refer to principal and subsidiary main positions and cycles. The distinction is only made when needed. In a principal main position a convex polygon flexagon has the appearance of four leaves each with a vertex at the centre so there are four pats and two sectors. Some are twisted bands and so exist as enantiomorphic (mirror image) pairs.

If a flexagon is regarded as a linkage then bending the leaves during flexing is not permissible. However, allowing bending during flexing does makes it easier to rationalise dynamic behaviours of the convex polygon flexagon family, and also makes the manipulation of some types of convex polygon flexagon more interesting.

The first variety of the convex polygon flexagon family, the digon flexagon, can only be flexed with the leaves truncated and then only by bending the leaves of a paper model: the 'push through' flex. The digon flexagon doesn't have a clearly defined intermediate position. For all other varieties of convex polygon flexagon an intermediate position is flat, and has the appearance of two polygons with a common side. The second and third varieties of the family are triangle flexagons and square flexagons. Triangle flexagons can be made to traverse complete cycles by truncating the leaves of paper models to regular hexagons and using the push through flex. Truncating the leaves of paper models of square flexagons to regular octagons makes it possible to traverse subsidiary 2-cycles by using push through flexes.

None of the first three varieties is typical of the convex polygon flexagon family. Some typical convex polygon flexagons are described in the next chapter.

7.1   The convex polygon family

There is an infinite family of convex polygon flexagons (Conrad 1960, Conrad and Hartline 1962, McIntosh 2000c, McIntosh 2000e, McIntosh 2000f). The leaves are regular convex polygons. Varieties are named after the constituent polygons. A feature of some varieties of convex polygon flexagon is that there may be more than one type of main position and more than one type of complete cycle. It then becomes necessary to refer to principal and subsidiary main positions and cycles. The distinction is only made when needed. The use of a numerical prefix to indicate the number of main positions in a cycle is particularly convenient for convex polygon flexagons. Thus, a 3-cycle is a cycle in which there are three main positions. All the main positions in a cycle always have the same appearance. The appearance of an intermediate position of a particular variety of a convex polygon flexagon is always the same, with the exception of the digon flexagon which doesn't have a clearly defined intermediate position.

An ideal convex polygon flexagon consists of a band of rigid regular convex polygons, with s sides, hinged together at common sides. A principal main position has the appearance of four ($2n$) polygons arranged about a point, each with a vertex at the point. It has twofold rotational symmetry and consists of two sectors. In general a principal main position of a convex polygon flexagon is not flat. The number of faces on a single complete principal cycle convex polygon flexagon is always the same as the number of sides on the constituent polygons, as is the number of main positions in the principal cycle. The number of possible arrangements of the leaves visible on a face is $2s$. Main position links between cycles are possible for $s > 2$. Box position links are possible for $s > 3$. There are various restrictions on the types of cycles that can be linked. For a given number of faces the number of distinct types of convex polygon flexagon increases as s increases.

If a flexagon is regarded as a linkage (Section 3.3) then bending the leaves during flexing is not permissible. However, allowing bending during flexing does makes it easier to rationalise dynamic behaviour of the convex polygon flexagon family, and does make the manipulation of some types of flexagon more interesting. Whether or not bending leaves during flexing is regarded as acceptable is a matter of personal taste. As the number of sides, s, on the polygons of a convex polygon flexagon increases, the dynamic behaviour becomes more complicated, and paper models become more difficult to handle. It is possible to make workable paper models for $s \leq 8$. Some properties of the single principal cycle types of

Table 7.1 *Single principal cycle types of the first nine varieties of convex polygon flexagons.*

Flexagon variety	Vertex angle	Regular flexagons Flexagon symbols	Numbers of distinct irregular flexagons	Numbers of distinct incomplete flexagons			
				Three faces	Four faces	Five faces	Six faces
Digon	$0°$	$\langle 2, 2 \rangle^a$	0	0	0	0	0
Triangle	$60°$	$\langle 3, 3 \rangle^b$	0	0	0	0	0
Square	$90°$	$\langle 4, 4 \rangle$	1	1	0	0	0
Pentagon	$108°$	$\langle 5, 5 \rangle$ $\langle 5, 5/2 \rangle$	2	1	3	0	0
Hexagon	$120°$	$\langle 6, 6 \rangle$	11	1	3	8	0
Heptagon	$128° \; 34'$	$\langle 7, 7 \rangle$ $\langle 7, 7/2 \rangle$ $\langle 7, 7/3 \rangle$	40^c	1	3	8	38
Octagon	$135°$	$\langle 8, 8 \rangle$ $\langle 8, 8/3 \rangle$		1	3	8	38
Enneagon	$140°$	$\langle 9, 9 \rangle$ $\langle 9, 9/2 \rangle$ $\langle 9, 9/4 \rangle$		1	3	8	38
Decagon	$144°$	$\langle 10, 10 \rangle$ $\langle 10, 10/3 \rangle$		1	3	8	38

[a] Cannot be flexed without disconnecting a hinge, refolding the flexagon, and reconnecting the hinge.

[b] A complete cycle cannot be traversed without disconnecting a hinge, refolding the flexagon, and reconnecting the hinge.

[c] Conrad and Hartline (1962)

the first nine varieties of the convex polygon family are summarised in Tables 7.1 and 7.2.

7.2   Degrees of freedom of intermediate positions

The first variety of the convex polygon flexagon family, the digon flexagon, doesn't have a clearly defined intermediate position. For all other varieties of convex polygon flexagon an intermediate position is flat, and has the appearance of two polygons with a common side (Fig. 7.1). In an intermediate position a convex polygon flexagon has a maximum of s degrees of freedom. These are summarised below for six varieties of the convex

Table 7.2 *Numbers and types of subsidiary cycles for the first nine varieties of single complete principal cycle convex polygon flexagons.*

Flexagon variety	2-Cycle	3-Cycle	4-Cycle	5-Cycle	6-Cycle	7-Cycle	8-Cycle	9-Cycle	10-Cycle
Digon	0	0	0	0	0	0	0	0	0
Triangle	0	0	0	0	0	0	0	0	0
Square	2	0	0	0	0	0	0	0	0
Pentagon	0	0	0	1	0	0	0	0	0
Hexagon	3	2	0	0	0	0	0	0	0
Heptagon	0	0	0	0	0	2	0	0	0
Octagon	4	0	2	0	0	0	1	0	0
Enneagon	0	3	0	0	0	0	0	2	0
Decagon	5	0	0	4	0	0	0	0	1

polygon family. There is a minimum of two degrees of freedom. It is always possible to rotate about hinges A (these are shown in Fig. 7.1) and it is always possible to flex to a principal main position.

(a) Triangle flexagons. It may be possible to flex to main positions using either hinges A and B, or hinges A and C (Fig. 7.1(a)).

(b) Square flexagons. It may be possible to flex to main positions using either hinges A and B, or hinges A and D (Fig. 7.1(b)). It may also be possible to open it to a box position using hinges A and C.

(c) Pentagon flexagons. It may be possible to flex to principal main positions using either hinges A and B, or hinges A and E (Fig. 7.1(c)). It may also be possible to open it into an open ended, oblique sided, box: an 'oblique box position' using either hinges A and C, or hinges A and D. An oblique box position may also be a subsidiary main position.

(d) Hexagon flexagons. It may be possible to flex to principal main positions using either hinges A and B, or hinges A and F (Fig. 7.1(d)). It may also be possible to open it to an oblique box position using either hinges A and C, or hinges A and E. Further, it may be possible to open it to a box position using hinges A and D. Both oblique box positions and box positions may also be subsidiary main positions.

(e) Heptagon flexagons. It may be possible to flex to principal main positions using either hinges A and B, or hinges A and G (Fig. 7.1(e)). It may also be possible to open it to an oblique box position using either hinges A and C, or hinges A and F. Further, it may be possible to open it to a different type of oblique box position using either hinges A and D or hinges A and E. The oblique box positions may also be subsidiary main positions.

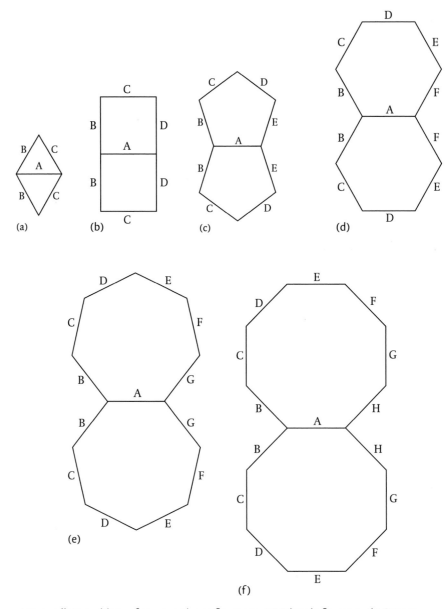

Fig. 7.1	Intermediate positions of convex polygon flexagons. (a) Triangle flexagon. (b) Square flexagon. (c) Pentagon flexagon. (d) Hexagon flexagon. (e) Heptagon flexagon. (f) Octagon flexagon.

(f) Octagon flexagons. It may be possible to flex to principal main positions using either hinges A and B, or hinges A and H (Fig. 7.1(f)). It may also be possible to open it into a flat ring of four octagons. Further, it may be possible to open the octagon flexagon to an oblique box position using either hinges A and D, or hinges A and F. Finally, it may be possible to

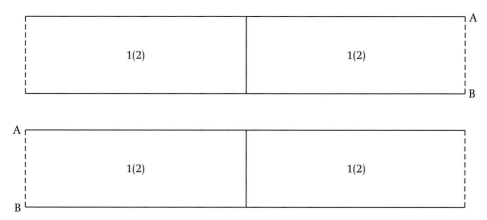

Fig. 7.2   Net for the truncated digon flexagon. Join the two parts of the net at A-B. Fold into an open ended box with the numbers 1 on the outside and join the ends.

Fig. 7.3   Tuckerman diagram for the truncated digon flexagon.

open it to a box position using hinges A and E. A flat ring of four octagons, and the box and oblique box positions, may all also be subsidiary main positions.

7.3   The digon flexagon

The digon flexagon is the first variety of the convex polygon family, and there is only one type. A digon (Coxeter 1963, Taylor 1997) is a regular polygon which has two vertices and two parallel sides of equal length. If the sides are of finite length then the sides are distinct but coincident and the digon is of zero width. If the sides are of infinite length then the digon may have nonzero width. An ideal digon flexagon is made from four infinitely long digons of nonzero width. It is an untwisted band so there is only one form. A main position has the appearance of a square tube of infinite length. To flex it the tube is turned inside out. This is only possible by disconnecting a hinge, refolding the flexagon, and reconnecting the hinge. The digon flexagon is of little intrinsic interest but it is included because it is needed for a systematic treatment of some of the higher varieties of the convex polygon flexagon family.

7.3.1   The truncated digon flexagon

The net for the truncated digon flexagon is shown in Fig. 7.2. The digons have been truncated to narrow rectangles. A main position has the

Fig. 7.4   Flexagon figure for the truncated digon flexagon.

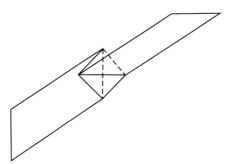

Fig. 7.5   Paper model of Hooke's universal joint.

appearance of an open ended box, similar to a square flexagon box position (Fig. 3.10) but shallower. As assembled the truncated digon flexagon is in main position 1(2). The number in the brackets is visible inside the box. To flex to the other main position, 2(1), the truncated digon flexagon has to be turned inside out. This can only be done by bending the leaves, which is easily done on the paper model by using a 'push through' flex in which the leaves are rotated about imaginary lines joining the hinges. Repeating the push through flex returns the flexagon to main position 1(2). Hence it is possible to traverse a 2-cycle. This cycle can be repeated indefinitely. The effect is that, as with the trihexaflexagon (Section 1.2) and also with other types of flexagon with complete cycles, the truncated digon flexagon is continually turned inside out. The cycle can also be traversed in the reverse direction. The Tuckerman diagram is shown in Fig. 7.3 and the flexagon figure in Fig. 7.4. In the figure the sides of the inscribed digon have been separated for clarity. The flexagon symbol is ⟨2, 2⟩.

7.3.2   The double hinged truncated digon flexagon

A variant of the truncated digon flexagon which can be flexed without bending the leaves may be made by connecting the leaves with a double hinge in the form of Hooke's universal joint (Dunkerley 1910). How this may be achieved in a paper model is shown in Fig. 7.5. The central part of the double hinge is a square and this is attached to the leaves using transparent adhesive tape. It is best to use tape on both sides of a leaf.

A main position of the double hinged truncated digon flexagon is flat and its appearance is shown in Fig. 7.6. The easiest way to make a paper

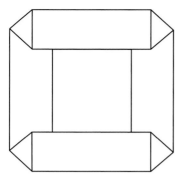

Fig. 7.6   Main position of the double hinged truncated digon flexagon.

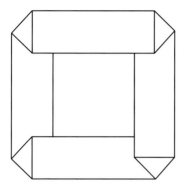

Fig. 7.7   Main position of a hybrid flexagon.

model is to cut out four leaves and four hinge squares and assemble them using the figure as a guide. The flexagon may be flexed by turning the leaves over in pairs. Doing this four times returns the flexagon to its initial configuration.

If one end of the flexagon is given a half twist before making the final connection it becomes a hybrid flexagon (Engel, 1969), and a main position has the appearance shown in Fig. 7.7. It is flexed by turning over leaves one at a time. Doing this eight times returns the flexagon to its initial configuration. It is a twisted band so exists in two enantiomorphic (mirror image) forms.

7.3.3   The flexatube

If the leaves of a digon flexagon are truncated to squares then a main position has the appearance of an open ended box, identical to that of a square flexagon box position (Fig. 3.10). It isn't then possible to flex a paper model by bending the leaves. However, the addition of diagonal hinges makes it possible to flex a paper model. The resulting structure is called a flexatube (Conrad and Hartline 1962, Gardner 1966, McIntosh 2000c, Mitchell 1999).

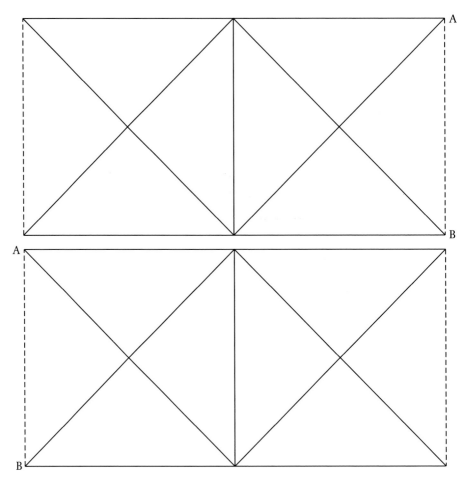

Fig. 7.8   Net for a flexatube. Join the two parts of the net at A-B. Crease all the lines. Fold into an open ended box and join the ends. Difficult.

The net for the flexatube is shown in Fig. 7.8. It is an untwisted band so there is only one form. It can be continually turned inside out using a complicated sequence of flexes. The flexatube is usually presented as a puzzle with the challenge: turn it inside out. There are several ways of doing this. The simplest sequence of flexes (Mitchell 1999) is shown in Fig. 7.9.

7.4   Triangle flexagons revisited

Triangle flexagons (Section 5.1) are the second variety of the convex polygon flexagon family. Main positions have the appearance shown in Fig. 3.4.

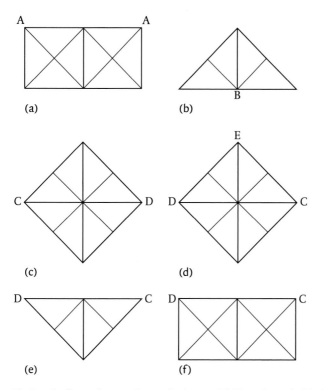

Fig. 7.9   Flexing the flexatube. (a) Flatten the box and fold vertices A behind to form a triangle. (b) Open the pocket at B, ease out the flap and fold down to form a square. (c) Turn over left to right, interchanging the positions of vertices C and D. (d) Fold the triangle CDE forwards and tuck into the pocket. (e) Fold the flaps at the back of the triangle downwards to form a flattened box. (f) Open up the flattened box. It is now inside out.

A cycle of a triangle flexagon cannot be traversed completely without disconnecting a hinge, refolding the flexagon, and reconnecting the hinge, even if bending the leaves is allowed.

7.4.1   The single cycle truncated triangle flexagon

Fig. 7.10 shows the net for the single cycle triangle flexagon (Fig. 5.1), but with the triangles truncated to regular hexagons. A main position has the appearance of a slant ring of four hexagons with a square hole at the centre, as shown in Fig. 7.11. The shorter hinges make it possible to bend the leaves of a paper model and use a push through flex to reverse the inside and outside of a main position. This makes it possible to traverse the cycle completely. Dashed lines are used on the intermediate position map (Fig. 7.12) to represent main positions. These indicate that bending the leaves is needed to traverse the cycle.

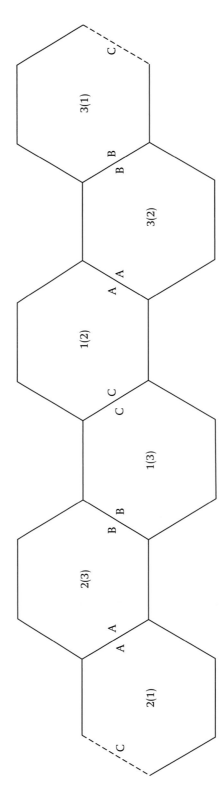

Fig. 7.10 Net for the single cycle truncated triangle flexagon. Fold leaves numbered 3 together, then leaves numbered 1, and join the ends.

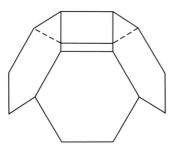

Fig. 7.11 Appearance of a main position of the single cycle truncated triangle flexagon.

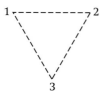

Fig. 7.12 Intermediate position map of the single cycle truncated triangle flexagon.

7.5 Square flexagons revisited

Square flexagons (Chapter 6) are the third variety of the convex polygon family. The full map of the regular single cycle square flexagon (Fig. 6.2) shows that all four intermediate positions can be opened up into a box position (Fig. 3.10). These intermediate positions occur in pairs with the codes 1|3], 3|1] and 2|4], 4|2]. If it were possible to turn the box positions inside out they could be regarded as subsidiary main positions with codes 1(3), 3(1) and 2(4), 4(2). It would then be possible to flex directly between the pairs of intermediate positions. However, it is not possible to turn the box positions of a paper model inside out even if bending the leaves is allowed (Subsection 7.3.3).

7.5.1 The regular single cycle truncated square flexagon

Fig. 7.13 shows the net for the regular single cycle square flexagon (Fig. 6.1), but with the squares truncated to regular octagons. A main position has the appearance of a flat ring of four octagons (Fig. 7.14). The short hinges make it possible to bend the leaves of a paper model and use a push through flex to turn a box position inside out and it becomes a subsidiary main position (Fig. 7.15). The cycle of the regular single cycle square flexagon (Subsection 6.1.2) becomes the principal 4-cycle of the regular single cycle truncated square flexagon. In addition there are two subsidiary 2-cycles in which the box positions of the regular single cycle square flexagon become subsidiary main positions. In the subsidiary 2-cycles the dynamic behaviour is the same as that of the truncated digon flexagon (Subsection 7.3.1).

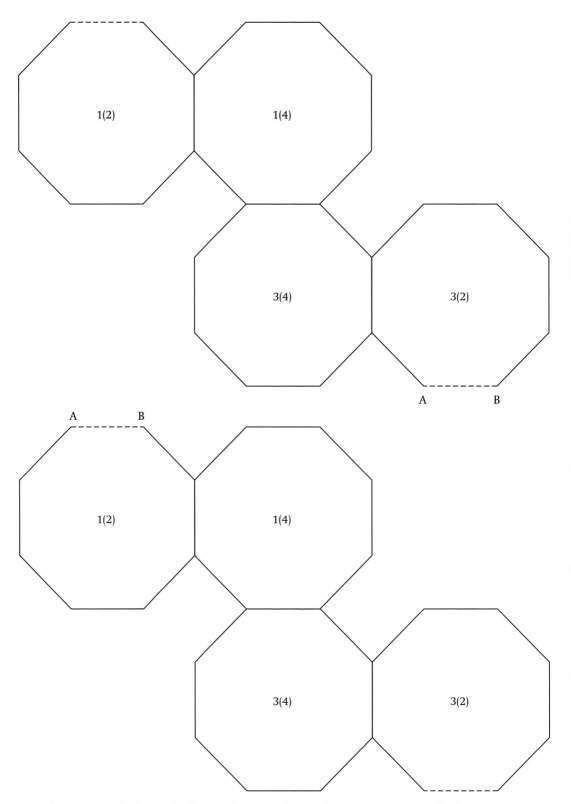

Fig. 7.13 Net for the regular single cycle truncated square flexagon. Join at A-B. Difficult.

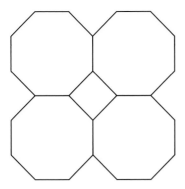

Fig. 7.14 Flat ring of four octagons.

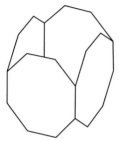

Fig. 7.15 Appearance of a subsidiary main position of the regular single principal cycle truncated square flexagon.

Fig. 7.16 Intermediate position map of the regular single principal cycle truncated square flexagon.

All three cycles are shown on the intermediate position map (Fig. 7.16). The face numbers have been chosen so that the principal 4-cycle appears as a square and the two 2-cycles as its diagonals. Starting at any intermediate position it is possible to flex directly to any other intermediate position. If bending of the leaves is not permitted then it is only possible to traverse the principal 4-cycle and the intermediate position map becomes as shown in Fig. 3.13.

The method of analysis does not distinguish between different types of single complete cycle flexagons so it also applies to the irregular single cycle square flexagon (Subsection 3.4.1) with the squares truncated to regular octagons.

8   Typical convex polygon flexagons

None of the first three varieties of convex polygon flexagon, described in the previous chapter, is typical of the family. The fourth variety, pentagon flexagons, and higher varieties have characteristics in common and all can be regarded as typical members of the family. Some are twisted bands and so exist as enantiomorphic (mirror image) pairs.

In a typical convex polygon flexagon the sum of the leaf vertex angles at the centre of a principal main position is greater than $360°$ so the principal main position is skew and its outline is a skew polygon. For a typical convex polygon flexagon with a single complete principal cycle the number of main positions in the principal cycle is the same as the number of sides on the constituent polygons. It is always possible to traverse the principal cycle without bending the leaves of paper models. There is always at least one subsidiary cycle. In general subsidiary cycles cannot be traversed without bending leaves. The appearance of the subsidiary main positions is different from that of the principal main positions. The number of main positions in a subsidiary cycle is either the same as the number in the principal cycle, or a factor of this number. Where the numbers of main positions in subsidiary cycles differ, so do the appearances of associated subsidiary main positions. When bending of leaves is allowed it is possible to traverse from a given intermediate position to any other intermediate position via just one main position.

In a principal main position a typical convex polygon flexagon has two possible configurations. To change from one configuration to the other peak folds become valley folds and valley folds become peak folds. A paper model may be snapped through from one configuration to the other by bending the leaves. This transformation is called a 'snap flex'. If a typical convex polygon flexagon contains a principal main position link between two cycles, then a snap flex is needed to traverse from one cycle to the other.

Various features of typical convex polygon flexagons are illustrated through descriptions of pentagon flexagons, hexagon flexagons and octagon flexagons. With octagon flexagons an additional type of flex, the 'twist flex', appears. When using the twist flex to traverse from one main position to another there is no clearly defined intermediate position. Some data are given on numbers of distinct types of pentagon flexagons and hexagon flexagons.

8.1 Characteristics of typical convex polygon flexagons

The first three varieties of convex polygon flexagon, the digon flexagon (Section 7.3), triangle flexagons (Section 7.4) and square flexagons (Section 7.5) are not typical of convex polygon flexagons. Analysis of typical convex polygon flexagons is less complete than it is for the first three varieties (Conrad 1960, Conrad and Hartline 1962, McIntosh 2000c, McIntosh 2000e, McIntosh 2000f).

More than one type of cycle is possible, and it is necessary to distinguish between principal and subsidiary cycles and main positions. A principal main position of a typical convex polygon flexagon has an appearance characteristic of the particular variety of flexagon. The sum of the leaf vertex angles at the centre of a principal main position is greater than 360°, so a principal main position is skew and the outline is a skew polygon. As an example Fig. 3.3 shows the appearance of a principal main position of a pentagon flexagon.

If a typical convex polygon flexagon is flexed from one intermediate position to an adjacent intermediate position on the principal cycle it moves in a continuous path. A principal main position is midway between the two intermediate positions. Both pairs of hinges connecting the pats move throughout the flex. By contrast, when flexing a square flexagon from one intermediate position to the next (Subsection 3.4.1) the two pairs of hinges move in sequence, one pair moves when flexing from an intermediate position to a main position, then the other pair moves when flexing from the main position to another intermediate position. This difference in dynamic behaviour is due to the effect of the angle excess over 360°, at the centre of a principal main position of a typical convex polygon flexagon, on the movement of hinges.

8.1.1 Single complete principal cycle convex polygon flexagons

For a typical convex polygon flexagon with a single complete principal cycle the number of main positions in the principal cycle is the same as the number of sides on the constituent polygons. It is always possible to traverse the principal cycle without bending the leaves.

There is always at least one subsidiary cycle. In general, subsidiary cycles cannot be traversed without bending the leaves. The appearance of the subsidiary main positions is different from that of the principal main positions. The numbers of main positions in a subsidiary cycle is either the same as the number in the principal cycle, or a factor of this number. Where the numbers of main positions in subsidiary cycles differ, so do the appearances of associated subsidiary main positions. When bending of leaves is allowed it is possible to traverse from a given intermediate position to any other intermediate via just one main position.

It is possible to draw an intermediate position map and then pick out the numbers and types of subsidiary cycles by finding all the regular polygons, including digons, connecting all possible pairs of intermediate positions. Numbers and types of subsidiary cycles for the first nine varieties of single complete principal cycle convex polygon flexagons are listed in Table 7.2.

In the presence of subsidiary cycles classification of typical convex polygon flexagons on the basis of the appearance of main positions needs care to avoid ambiguity. Classification on the basis of the appearance of the intermediate positions shown in Fig. 7.1 is unambiguous, but less helpful.

8.1.2   The snap flex

In a principal main position a typical convex polygon flexagon has two possible configurations. To change from one configuration to the other peak folds become valley folds and valley folds become peak folds. In Fig. 3.3 the initial peak folds run left to right and the initial valley folds top to bottom. A paper model may be snapped through from one configuration to the other by bending the leaves. This transformation is called a 'snap flex'. If the leaves are rigid then the change can only be made by disconnecting a hinge, refolding the convex polygon flexagon, and reconnecting the hinge. If a typical convex polygon flexagon contains a principal main position link, then a snap flex is needed to traverse from one cycle to the other.

8.2   Pentagon flexagons

Pentagon flexagons are the fourth variety of the convex flexagon family. Paper models of pentagon flexagons work well. The continuous path when flexing between adjacent intermediate positions on a principal cycle is aesthetically satisfying.

8.2.1   Types of single principal cycle pentagon flexagon

Flexagon figures for all possible types of single principal cycle pentagon flexagon are shown in Fig. 8.1. These data are for the principal cycle. The data shown in Figs. 8.1(a) and 8.1(b) agree with data given by Conrad and Hartline (1962). The flexagon figures were derived by considering possible polygons joining midpoints of the sides of the circumscribing pentagon, and then checked by making paper models using the method described

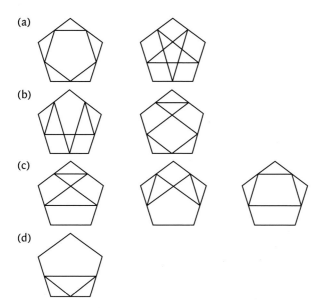

Fig. 8.1 Flexagon figures for all possible single principal cycle pentagon flexagons. (a) Regular. (b) Irregular. (c) Incomplete, four faces. (d) Incomplete, three faces.

in Section 6.2. During these checks it was found that for an incomplete single principal cycle pentagon flexagon the flexagon figure must connect a group of adjacent sides of the circumscribing pentagon. This procedure established confidence in the use of flexagon figures to establish possible types of single principal cycle convex polygon flexagons and it was used to derive the data shown in Table 7.1, except where noted.

There are two types of regular single principal cycle pentagon flexagon (Fig. 8.1(a)). This is because there are two types of regular pentagon: the regular convex polygon, and the regular star pentagon which is obtained by stellating the convex pentagon. There are three incomplete types with four faces (Fig. 8.1(c)) and there is one incomplete type with three faces (Fig. 8.1(d)).

8.2.2 The regular single principal cycle pentagon flexagon ⟨5, 5⟩

Fig. 8.2 shows the net for the regular single principal cycle pentagon flexagon ⟨5, 5⟩. This is one of the two regular single principal cycle pentagon flexagons. The flexagon figure is shown in Fig. 8.1(a), left. The simplified map for the principal cycle is shown in Fig. 8.3. This cycle can be traversed without bending the leaves. The intermediate position map is shown in Fig. 8.4.

As assembled the pentagon flexagon is in intermediate position 1[3, 4] (Fig. 8.3). There are two numbers in the square bracket because the

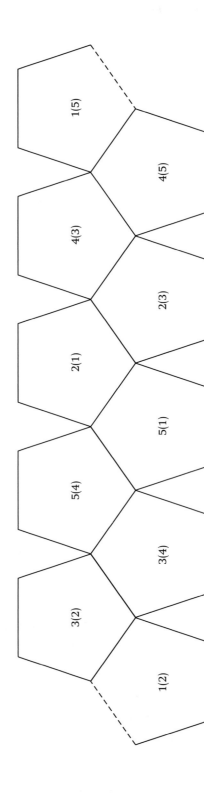

Fig. 8.2 Net for the regular single principal cycle pentagon flexagon (5, 5). Fold until only face number 1 is visible. Difficult.

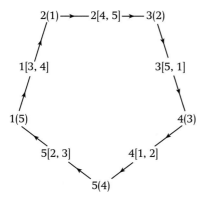

Fig. 8.3   Simplified map of the principal cycle of the regular single principal cycle pentagon flexagon ⟨5, 5⟩.

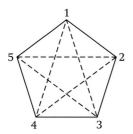

Fig. 8.4   Intermediate position map of the regular single principal cycle pentagon flexagon ⟨5, 5⟩.

intermediate position can be opened up into an oblique open ended box in two different ways. This is true of all the intermediate positions. These open ended boxes are subsidiary main positions of a subsidiary 5-cycle, shown by dashed lines in Fig. 8.4. These subsidiary main positions have the appearance of a ring of four pentagons with a square hole at the centre, similar to Fig. 7.11. If intermediate position 1[3, 4] is opened using appropriate hinges then face 3 is visible inside the oblique box so it is subsidiary main position 1(3). Carrying out a push through flex to subsidiary main position 3(1) and closing the oblique box leads to intermediate position 3[5, 1]. The subsidiary 5-cycle may be traversed by repeating the sequence. In the first repeat intermediate position 3[5, 1] is opened up so that face number 5 is visible. The above analysis does not distinguish between different types of complete single principal cycle convex polygon flexagons so it applies to all the flexagons whose flexagon figures are shown in Figs. 8.1(a) and 8.2(b).

At all principal main positions alternate pats are single leaves and folded piles of four leaves so principal main position links between principal cycles are possible. There are no single leaves at subsidiary main positions so subsidiary main position links between subsidiary cycles are not possible.

8.2.3   The regular single principal cycle pentagon flexagon ⟨5, 5/2⟩

Fig. 8.5 shows the net for the regular single principal cycle pentagon flexagon ⟨5, 5/2⟩. This is the second of the two regular single principal cycle pentagon flexagons. The flexagon figure is shown in Fig. 8.1(a), right. The faces have been numbered so that its simplified map and intermediate position map are identical to those shown in Figs. 8.3 and 8.4. The analysis is the same as that for the regular single principal cycle pentagon flexagon ⟨5, 5⟩ (Subsection 8.2.2).

At all subsidiary main positions alternate pats are single leaves and folded piles of four leaves so subsidiary main position links between subsidiary cycles are possible. These are analogous to box position links in square flexagons (Section 6.5) and occur in pairs. There are no single leaves at principal main positions so principal main position links between principal cycles are not possible.

8.2.4   Pentagon flexagon with two incomplete principal cycles and a principal main position link

Fig. 8.6 shows the net for a pentagon flexagon with two incomplete principal cycles and a principal main position link. The simplified map for the principal cycles (Fig. 8.7) shows that they are linked at main position 2(1). The diamond indicates the need for a snap flex at this main position to traverse from one cycle to the other. The intermediate position map is shown in Fig. 8.8). There are two pseudo 3-cycles each containing two main positions belonging to the principal cycle of a pentagon flexagon and one subsidiary main position belonging to the subsidiary cycle of a pentagon flexagon. The common main position, code 1(2), is a principal main position. Push through flexes are needed to completely traverse the pseudo 3-cycles, so the subsidiary main positions, codes 2(3) and 2(4), are shown by dashed lines to indicate the need to bend the leaves.

8.2.5   Pentagon flexagon with two incomplete principal cycles and a pair of subsidiary main position links

Fig. 8.9 shows the net for a pentagon flexagon with two incomplete principal cycles and a pair of subsidiary main position links. The face numbers and hinge letters have been chosen so that the simplified map for principal main positions is identical to that shown in Fig. 6.27 of square flexagons with two incomplete cycles and a pair of box position links

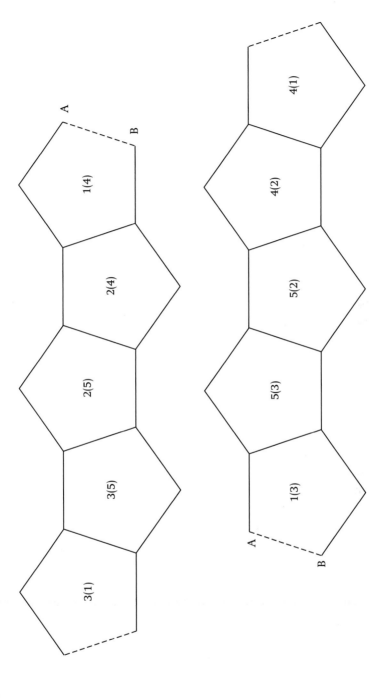

Fig. 8.5 Net for the regular single principal cycle pentagon flexagon (5, 5/2). Join the two parts at A-B. Fold until only face number 1 is visible. Difficult.

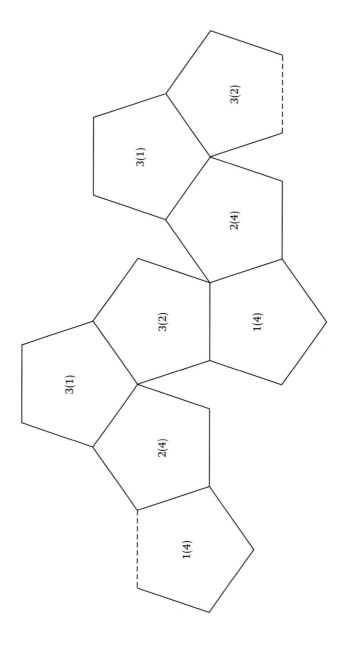

Fig. 8.6   Net for pentagon flexagon with two incomplete principal cycles, principal main position link. Fold until only face number 1 is visible.

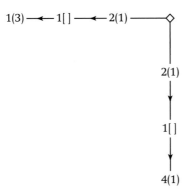

Fig. 8.7 Simplified map of pentagon flexagon with two incomplete principal cycles, principal main position link. Difficult.

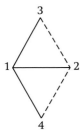

Fig. 8.8 Intermediate position map of pentagon flexagon with two incomplete principal cycles, principal main position link.

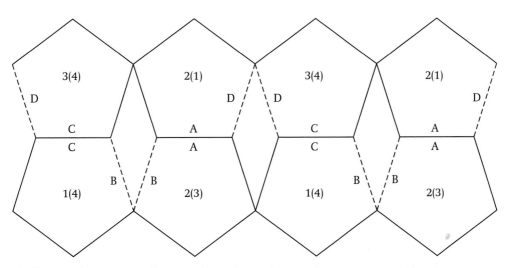

Fig. 8.9 Net for pentagon flexagon with two incomplete principal cycles, pair of subsidiary main position links. Join at adjacent dashed lines. Fold until only face number 1 is visible. Difficult.

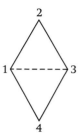

Fig. 8.10   Intermediate position map of pentagon flexagon with two incomplete principal cycles, pair of subsidiary main position links.

(Subsection 6.5.2). The net is closely related to that shown in Fig. 6.26. The pseudo-4-cycle shown in the figure can be traversed without bending the leaves.

The intermediate position map (Fig. 8.10) shows that there are two pseudo 3-cycles each containing two main positions belonging to the principal cycle of a regular single principal cycle pentagon flexagon and one subsidiary main position belonging to the subsidiary cycle of a regular single principal cycle pentagon flexagon. The common main position, code 1(3), is a subsidiary main position. Push through flexes are needed to completely traverse the pseudo 3-cycles so the subsidiary main position, code 1(3), is shown by a dashed line. The interpretation in terms of Fig. 6.27 is that, say, intermediate position 1[3] can be opened into an oblique open ended box which is subsidiary main position 1(3). A push through flex may then be used to convert it into subsidiary main position 3(1). The box is then closed to reach intermediate position 3[1].

8.3   Hexagon flexagons

Hexagon flexagons are the highest variety of convex polygon flexagon whose paper models are reasonably easy to handle. They should not be confused with hexaflexagons (Chapter 4), which are very different. The dynamic behaviour of hexagon flexagons is similar to that of pentagon flexagons, with similar possibilities for the formation of principal main position links and subsidiary main position links.

8.3.1   Types of single principal cycle hexagon flexagon
Flexagon figures for all possible types of single principal cycle hexagon flexagon are shown in Fig. 8.11. These data are for the principal cycle. The data in Fig. 8.11(a) agree with data given by Conrad and Hartline (1962). The flexagon figures were derived by considering possible polygons

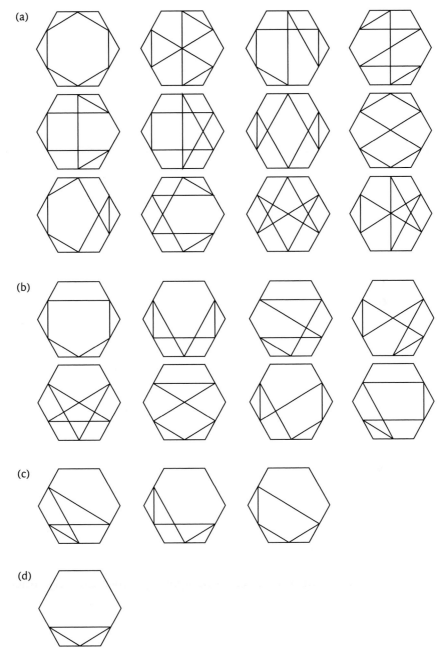

Fig. 8.11 Flexagon figures for all possible single principal cycle hexagon flexagons. (a) Regular (top left) and irregular. (b) Incomplete, five faces. (c) Incomplete, four faces. (d) Incomplete, three faces.

joining midpoints of the sides of the circumscribing hexagon, and then checked as described in Subsection 8.2.1.

There is only 1 type of regular single principal cycle hexagon flexagon (Fig. 8.11(a), top left). This is because stellating a regular hexagon results in two separate, intersecting triangles. There are 11 types of irregular single principal cycle hexagon flexagon, also shown in the figure. There are 8 types of incomplete single principal cycle hexagon flexagon with five faces (Fig. 8.11(b)), 3 incomplete types with four faces (Fig. 8.11(c), and 1 incomplete type with three faces (Fig. 8.11(d)).

8.3.2   The regular single principal cycle hexagon flexagon

Fig. 8.12 shows the net for the regular single principal cycle hexagon flexagon. The flexagon figure is shown in Fig. 8.11(a), top left. The simplified map for the principal cycle is shown in Fig. 8.13. This cycle can be traversed without bending the leaves.

As assembled the hexagon flexagon is in intermediate position 1[3, 4, 5] (Fig. 8.13). There are three numbers in the square brackets because the intermediate position can be opened up into an open ended box in one way, and into an oblique open ended box in two different ways. This is true of all the intermediate positions. An open ended box is a subsidiary main position of a subsidiary 2-cycle and has an appearance similar to that shown in Fig. 7.15. An oblique open ended box is a subsidiary main position of a subsidiary 3-cycle and has the appearance of a ring of four hexagons with a square hole at the centre, as shown in Fig. 7.11.

The intermediate position map is shown in Fig. 8.14. There are three subsidiary 2-cycles, shown by the three dashed lines passing through the centre of the intermediate position map. In these 2-cycles the dynamic behaviour of the hexagon flexagon is the same as that of the truncated digon flexagon (Subsection 7.3.1) and push through flexes are needed to traverse them. There are two subsidiary 3-cycles. In these 3-cycles the dynamic behaviour is the same as that of the truncated single cycle triangle flexagon (Subsection 7.4.1) and push through flexes are needed to traverse them. In one of the subsidiary 3-cycles the face numbers are odd, and in the other they are even. The above analysis does not distinguish between different types of complete single principal cycle convex polygon flexagon so it applies to all the flexagons whose flexagon figures are shown in Fig. 8.11(a).

At all principal main positions alternate pats are single leaves and folded piles of five leaves so principal main position links are possible. There are no single leaves at subsidiary main positions so subsidiary main position links are not possible.

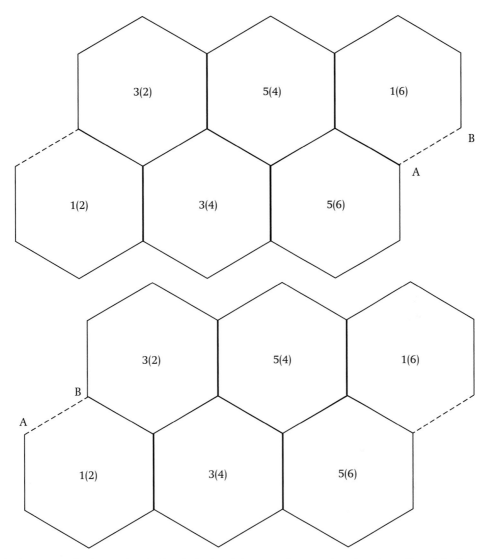

Fig. 8.12 Net for the regular single principal cycle hexagon flexagon. Cut along heavy lines. Join the two parts at A-B. Fold until only face number 1 is visible. Difficult.

8.3.3 An irregular single principal cycle hexagon flexagon

Fig. 8.15 shows the net for one of the irregular single principal cycle hexagon flexagons. This particular hexagon flexagon is a good example of an irregular single principal cycle convex polygon flexagon. Its irregular behaviour while traversing its principal cycle is clearly noticeable. The flexagon figure is shown in Fig. 8.11(a), top right. The faces have been numbered so that its simplified map and intermediate position map are the same as those shown in Figs. 8.13 and 8.14. The analysis is the same

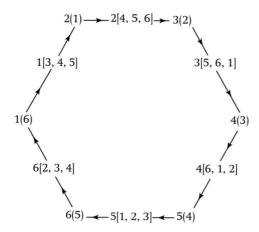

Fig. 8.13   Simplified map of the regular single principal cycle hexagon flexagon.

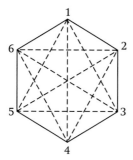

Fig. 8.14   Intermediate position map of the regular single principal cycle hexagon flexagon.

as that for the regular single principal cycle hexagon flexagon (Subsection 8.3.2).

At principal main positions 1(6) and 4(3) alternate pats are single leaves and folded piles of five leaves, so at these main positions principal main position links are possible. Pairs of subsidiary main position links are possible at subsidiary main positions 1(5) and 5(1), 2(4) and 4(2), 2(5) and 5(2), and 3(6) and 6(3).

8.4   Octagon flexagons

In general, paper models of octagon flexagons are difficult to handle because of the relatively short hinges and the large number of degrees of freedom. The dynamic behaviour of octagon flexagons is in general similar to that of pentagon flexagons and hexagon flexagons (Sections 8.2 and 8.3), with similar possibilities for the formation of principal main position links and subsidiary main position links. There are two additional

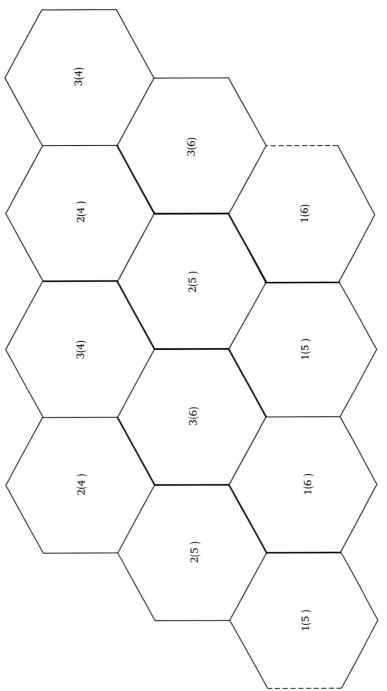

Fig. 8.15 Net for an irregular single principal cycle hexagon flexagon. Cut along heavy lines. Fold until only face number 1 is visible. Difficult.

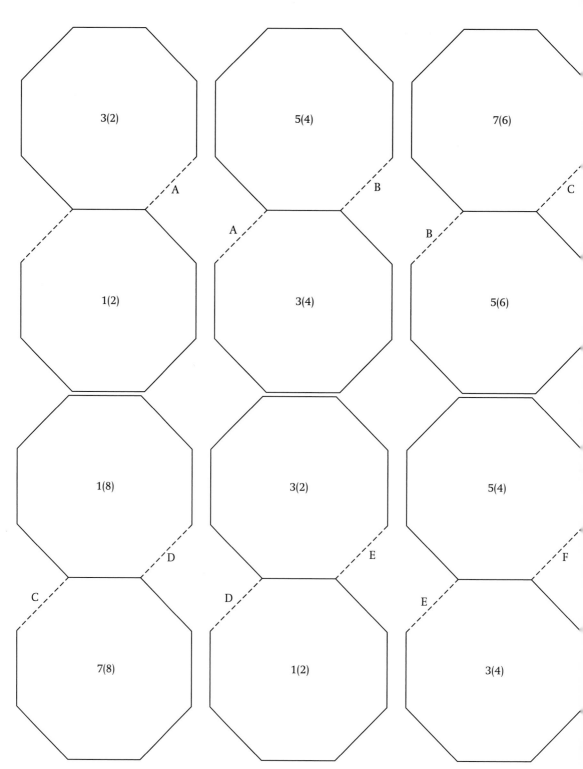

Fig. 8.16 Net for the regular single principal cycle octagon flexagon ⟨8, 8⟩. Join the eight parts at A, B, C, D, E, F and G, keeping the parts the same way up. Fold until only face number 1 is visible. Difficult. (Continued on next page)

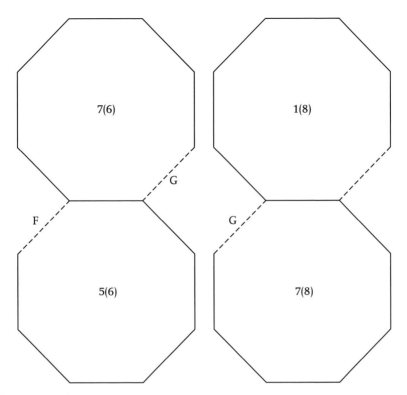

Fig. 8.16 (*Continued*)

features, which do not appear in lower varieties of the convex polygon flexagon family. Firstly, subsidiary cycles which may be traversed without bending the leaves are possible. Secondly, it is possible to use a 'twist flex', without bending the leaves, to traverse between some pairs of main positions without visiting an intermediate position. An intermediate position map does not show twist flex traverses.

The number of distinct types of single principal cycle octagon flexagons is large, and has not been completely determined (Table 7.1).

8.4.1 The regular single principal cycle octagon flexagon ⟨8, 8⟩

The net for the regular single principal cycle octagon flexagon ⟨8, 8⟩ is shown in Fig. 8.16. This is one of the two regular single principal cycle octagon flexagons. The flexagon figure is shown in Fig. 8.17. The simplified map for the principal cycle is shown in Fig. 8.18. The principal cycle can be traversed without bending the leaves.

The intermediate position map is shown in Fig. 8.19. There are four subsidiary 2-cycles, shown by the four dashed lines passing through the

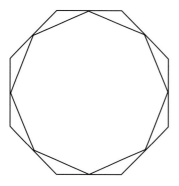

Fig. 8.17   Flexagon figure for the regular single principal cycle octagon flexagon ⟨8, 8⟩.

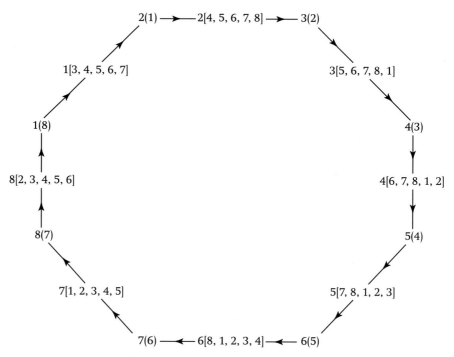

Fig. 8.18   Simplified map of the principal cycle of the regular single principal cycle octagon flexagon ⟨8, 8⟩.

centre of the intermediate position map. In these 2-cycles the dynamic behaviour of the regular single principal cycle octagon flexagon is the same as that of the truncated digon flexagon (Subsection 7.3.1) and push through flexes are needed to traverse them. There are two subsidiary 4-cycles. In these 4-cycles the dynamic behaviour is the same as that of the regular single cycle truncated square flexagon (Subsection 7.5.1) and the 4-cycles can be traversed without bending the leaves. In one of the subsidiary 4-cycles the face numbers are odd, and in the other they are

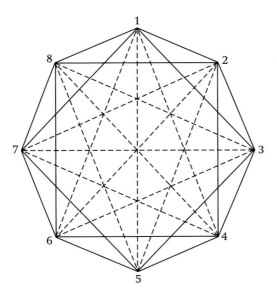

Fig. 8.19 Intermediate position map of the regular single principal cycle octagon flexagon ⟨8, 8⟩.

even. There is one subsidiary 8-cycle and push through flexes are needed to traverse it.

As assembled the regular single principal cycle octagon flexagon is in intermediate position 1[3, 4, 5, 6, 7]. There are five numbers in the square brackets because the intermediate position can be opened up into an open ended box in one way, into an oblique open ended box in two different ways, and into a flat ring of four octagons in two different ways. This is true of all the intermediate positions. An open ended box is a subsidiary main position of a subsidiary 2-cycle and has the appearance shown in Fig. 7.15. An oblique open ended box is a subsidiary main position of the subsidiary 8-cycle and has the appearance of a ring of four octagons with an irregular eight sided hole at the centre, rather similar to Fig. 7.11. A flat ring of four octagons is a subsidiary main position of a subsidiary 4-cycle and has the appearance shown in Fig. 7.14.

At all principal main positions alternate pats are single leaves and folded piles of seven leaves so principal main position links are possible. No subsidiary main position links are possible.

8.4.2 The three faced octagon flexagon
Fig. 8.20 shows the net for the three faced octagon flexagon. This is the simplest octagon flexagon and is reasonably easy to handle. The flexagon figure is shown in Fig. 8.21. The three faced octagon flexagon is a remarkable flexagon. It is a linkage with only six links, the same as Sarrut's

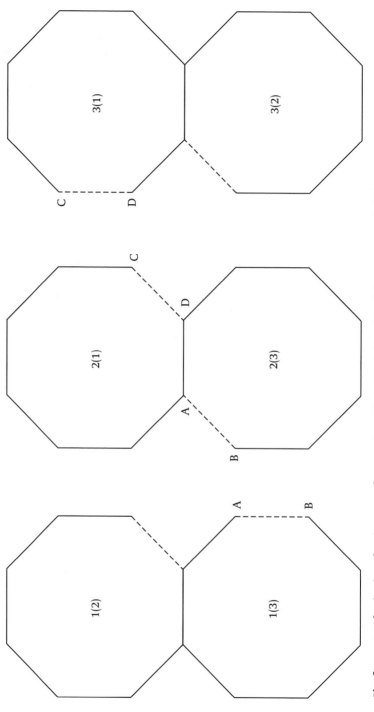

Fig. 8.20 Net for the three faced octagon flexagon. Join at A-B, C-D. Fold until only face number 2 is visible.

Fig. 8.21 Flexagon figure for the three faced octagon flexagon.

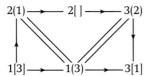

Fig. 8.22 Simplified map of the three faced octagon flexagon.

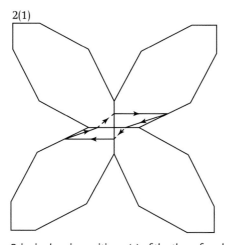

Fig. 8.23 Principal main position 2(1) of the three faced octagon flexagon, projected onto a plane.

parallel motion (Subsection 3.3.1) and three fewer than the trihexaflexagon (Sections 1.1 and 4.2). Nevertheless, its dynamic behaviour is more complicated than that of the trihexaflexagon, and this is achieved without bending the leaves.

The simplified map is shown in Fig. 8.22. The double lines show transformations which are possible by using the twist flex, described below. Main positions 2(1) and 3(2) are equivalent to principal main positions of the regular single principal cycle octagon flexagon ⟨8, 8⟩ (Subsection 8.4.1) and are therefore principal main positions. Similarly, main position 1(3) is equivalent to a subsidiary main position of a subsidiary 4-cycle of the regular single principal cycle octagon flexagon ⟨8, 8⟩ and is a subsidiary main position. Principal main position 2(1) is shown in Fig. 8.23, projected onto a plane. Subsidiary main position 1(3) is shown in Fig. 8.24. Main

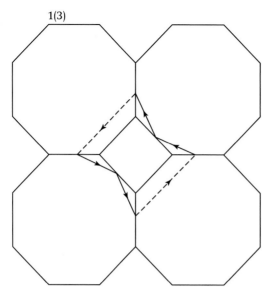

Fig. 8.24   Subsidiary main position 1(3) of the three faced octagon flexagon.

positions 2(1), 3(2) and 1(3) form a pseudo 3-cycle since it is made up from parts of two different cycles and the main positions do not all have the same appearance.

8.4.3   The twist flex

To twist flex the three faced octagon flexagon (Subsection 8.4.2) from principal main position 1(2) to subsidiary main position 1(3) hold the two pats which are single leaves and twist gently, allowing the other two pats to unfold. All six hinges move in an aesthetically satisfying manner. To reverse the flex, hold the top leaf in each pat of twofolded leaves and twist gently. Ideally, the leaves don't have to be bent. In a paper model bending is needed to make room for fingers. The front cover shows the flexagon the other way up moving away from subsidiary main position 3(1).

In theory the regular single principal cycle octagon flexagon ⟨8, 8⟩ (Subsection 8.4.1) can be twist flexed, but in practice this is difficult to achieve in a paper model.

9 Ring flexagons

In a systematic treatment flexagons can be classified into two main infinite families. The first is the convex polygon flexagon family and the second is the star flexagon family. A principal main position of a star flexagon is flat, and has the appearance of an even number of regular polygons arranged about its centre, each with a vertex at the centre. The first two varieties of star flexagons are square flexagons and hexaflexagons. Square flexagons are the first variety of the convex polygon flexagon family. The first two varieties are not typical of star flexagons. Typical star flexagons have at least eight polygons arranged about the centre of a principal main position, and the constituent polygons are regular star polygons. The dynamic behaviour of a typical star polygon flexagon is similar to that of the corresponding convex polygon flexagon. Interpenetration of the stellations during flexing doesn't matter in the ideal situation but does make the construction of paper models impossible.

Typical star flexagons are precursors to ring flexagons. If all the stellations are removed from the constituent polygons of a star polygon flexagon then it becomes a ring flexagon. Hence for any given star polygon flexagon there is always a corresponding ring flexagon with similar dynamic behaviour, and there is an infinite family of ring flexagons. A principal main position of a ring flexagon has the appearance of a flat ring of an even number of regular convex polygons. The rings are regular in that each polygon is the same distance from the centre of the ring. The relatively short length of the hinges, the large number of pats in a main position, and the large numbers of degrees of freedom, make paper models of ring flexagons awkward and tedious to handle. In practice it is usually only possible to flex them to all possible main positions by continual reference to an appropriate map.

A compound flexagon is a ring flexagon in which alternate pats lie closer to the centre of a main position than do the others. There is an infinite number of compound flexagon varieties. Principal main positions are flat and have the appearance of compound rings of regular convex polygons in which alternate polygons lie closer to the centre of the ring. The leaves are regular convex polygons, and compound flexagons are named after the constituent polygons. The lines of hinges between pats do not intersect at the centres of the rings. Because of this, compound flexagons can only be flexed by bending the leaves. Intermediate positions are not clearly defined. Flexing paper models is difficult. Subsidiary cycles and subsidiary main positions sometimes appear but they are usually difficult to find.

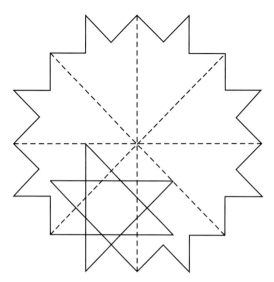

Fig. 9.1   A principal main position of a star octagon flexagon.

9.1   Star flexagons

Star flexagons are precursors to ring flexagons. There is an infinite family of star flexagons. The leaves are regular polygons. Varieties are named after the constituent polygons. An ideal star flexagon consists of a band of rigid regular polygons with s sides, hinged together at common sides. A principal main position is flat, and has the appearance of $2n$ polygons arranged about its centre, each with a vertex at the centre. It has n-fold rotational symmetry and consists of n sectors. An intermediate position has the appearance of n polygons, all with a common side. The first two varieties of star flexagons are square flexagons ($n = 2, s = 4$, Chapter 6) and hexaflexagons ($n = 3, s = 3$, Chapter 4). These are not typical of star flexagons.

Typical star flexagons have $n > 3$ and the constituent polygons are regular star polygons. As an example Fig. 9.1 shows the outline of a principal main position of a star octagon flexagon ($n = 4, s = 8$). The dashed lines show the positions of hinges between star octagons. For clarity only one of the eight star octagons is shown. Interpenetration of the stellations during flexing doesn't matter in the ideal situation but does make the construction of paper models impossible. Table 9.1 shows some details of the first six varieties of star flexagons, including flexagon symbols for all the regular single principal cycle star flexagons.

The dynamic behaviour of a typical star flexagon is similar to that of the corresponding convex polygon flexagon (Chapter 8). One difference is

Table 9.1 *First six varieties of star flexagons.*

Type of flexagon	n	s	Leaf vertex angle	Regular flexagons Flexagon symbols
Square	2	4	90°	⟨4, 4⟩
Hexaflexagon	3	3	60°	⟨3, 3⟩
Star octagon	4	8	45°	⟨8/3, 8⟩
				⟨8/3, 8/3⟩
Star pentagon	5	5	36°	⟨5/2, 5⟩
				⟨5/2, 5/2⟩
Star dodecagon	6	12	30°	⟨12/5, 12⟩
				⟨12/5, 12/5⟩
Star heptagon	7	7	25° 43′	⟨7/3, 7⟩
				⟨7/3, 7/2⟩
				⟨7/3, 7/3⟩

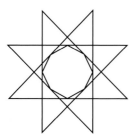

Fig. 9.2 Flexagon figure for the regular single principal cycle star octagon flexagon ⟨8/3, 8⟩.

that flexagon figures are inscribed in a regular star polygon rather than in a regular convex polygon. For example Fig. 9.2 shows the flexagon figure for the regular single principal cycle star octagon flexagon ⟨8/3, 8⟩.

9.2 The ring flexagon family

Ring flexagons are variants of typical star flexagons (Section 9.1). If all the stellations are removed from the constituent polygons of a star flexagon then it becomes a ring flexagon. Hence for any given star flexagon there is always a corresponding ring flexagon and there is an infinite family of ring flexagons as pointed out by Conrad and Hartline (1962). The leaves are regular convex polygons, and ring flexagons are named after the constituent polygons.

A principal main position of a ring flexagon has the appearance of a flat ring of $2n$ regular convex polygons with s sides. The rings are regular

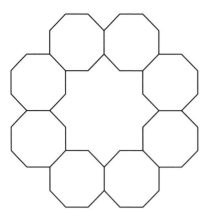

Fig. 9.3   Regular ring of eight octagons.

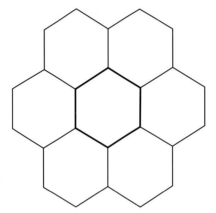

Fig. 9.4   Regular ring of six hexagons.

in that each polygon is the same distance from the centre of the ring. For example the appearance of a principal main position of a ring octagon flexagon is a ring of eight octagons (Fig. 9.3). Regular polygon rings have been discussed by several authors (Dunlap 1997/8, Griffiths 2001, Hirst 1995). An intermediate position of a ring flexagon has the appearance of n convex polygons all with a common side.

Some subsidiary positions of other varieties of flexagon have the appearance of flat rings of polygons. For example subsidiary main positions of subsidiary 4-cycles of octagon flexagons (Subsection 8.4.1) have the appearance of a ring of four octagons (Fig. 7.14). If the number of sectors in a hexagon flexagon (Section 8.3) is increased from two to three then subsidiary main positions of 3-cycles have the appearance of a ring of six hexagons (Fig. 9.4).

To move from a principal main position of a ring flexagon to an intermediate position pats are pinched together in n pairs in the first part of a pinch flex. In order to ensure correct flexing it is sufficient, but not always necessary, to restrict movements to those which have n-fold rotational symmetry. Imposing n-fold rotational symmetry during some flexes may make it necessary to bend the leaves. This can usually be avoided by relaxing the n-fold rotational symmetry requirement, but care is then needed to avoid spurious flexes. The dynamic behaviour of a ring polygon flexagon is similar to that of the corresponding convex polygon flexagon (Chapter 8). One difference is that the roles of principal and subsidiary cycles are sometimes interchanged.

The relatively short length of the hinges, the large number of pats in a main position, and the large numbers of degrees of freedom, make paper models of ring flexagons awkward and tedious to handle. It is easy to get them badly tangled. In practice it is usually only possible to flex them to all possible main positions by continual reference to an appropriate map.

9.2.1 A regular single principal cycle ring pentagon flexagon

Fig. 9.5 shows the net for the regular single principal cycle ring pentagon flexagon $\langle 5, 5 \rangle$. This corresponds to the regular single principal cycle pentagon flexagon $\langle 5, 5 \rangle$ (Subsection 8.2.2). A principal main position has the appearance of a ring of 10 pentagons (Fig. 9.6). The flexagon figure is the same as for the regular single principal cycle pentagon flexagon $\langle 5, 5 \rangle$ (Fig. 8.1(a), left). The intermediate position map is shown in Fig. 9.7. Both principal and subsidiary cycles can be traversed without bending the leaves so all the lines on the figure representing main positions are solid. Face numbers have been chosen so that the intermediate position map is otherwise the same as for the regular single principal cycle pentagon flexagon $\langle 5, 5 \rangle$ (Fig. 8.4). The roles of the principal and subsidiary cycles, both 5-cycles, are interchanged, and this is reflected in different face numbering sequences on the nets (Figs. 8.2 and 9.5).

9.3 Compound flexagons

A compound flexagon is a ring flexagon in which alternate pats lie closer to the centre of a main position than do the others. There is an infinite number of compound flexagon varieties. Flat main positions have the appearance of compound rings of regular convex polygons in which alternate polygons lie closer to the centre of the ring. The leaves are regular convex polygons, and compound flexagons are named after the constituent polygons. In some compound rings there is a slit rather than a hole at the centre of a ring. Some links between cycles are possible.

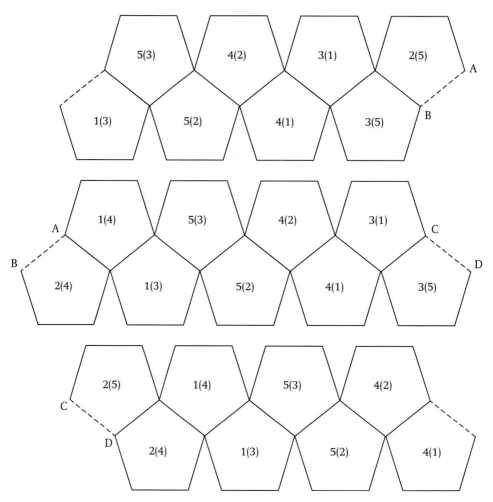

Fig. 9.5 Net for the regular single principal cycle ring pentagon flexagon ⟨5, 5⟩. Join the three parts of the net at A-B, C-D. Difficult.

Some possible flat main position appearances are given by Conrad and Hartline (1962) for compound flexagons made from squares, regular pentagons and regular hexagons.

  The lines of hinges between pats do not intersect at the centres of the rings. Because of this, compound flexagons can only be flexed by bending the leaves. Intermediate positions are not clearly defined, although for some types of compound flexagon they can be characterised by a single face number. Flexing paper models is difficult. Subsidiary cycles and subsidiary main positions sometimes appear but they can be difficult to find.

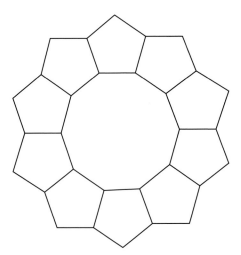

Fig. 9.6   Regular ring of 10 pentagons.

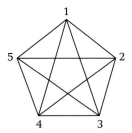

Fig. 9.7   Intermediate position map of the regular single principal cycle ring pentagon flexagon
$\langle 5, 5 \rangle$.

9.3.1   Modified flexagon figures for compound flexagons

The lack of clearly defined intermediate positions means that the flexagon
figures described in Section 3.5 cannot be used to characterise compound
flexagons. In a 'modified flexagon figure' part polygons are inscribed
in pairs of circumscribing polygons. A pair of circumscribing polygons
represents adjacent pats (one sector) in a flat main position. Dashed lines
show where the pats are attached to further pats. A circuit of line segments inscribed in the pair of circumscribing polygons represents a complete cycle. The circumscribing polygons have the same number of sides
as the appropriate leaves.

The part circuit of line segments inscribed in each circumscribing
polygon is part of a regular polygon. It starts and finishes on sides of a
circumscribing polygon where it is attached to the next pat. Solid lines
indicate that line segments pass upwards through a pat and dashed lines
that they pass downwards. The direction of the line segments remains the
same when passing from one circumscribing polygon to the other, but

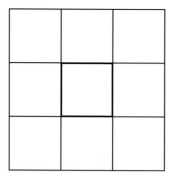

Fig. 9.8   Compound ring of eight squares.

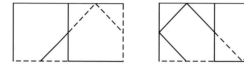

Fig. 9.9   Modified flexagon figures for both possible single complete cycle compound square flexagons.

the upwards or downwards direction changes. The total number of line segments in the circuit is the same as the number of main positions in the complete cycle. Modified flexagon figures occur in pairs in which the average number of main positions in a cycle is the same as the number of sides on the appropriate leaves.

Possible distinct types of single complete cycle compound flexagons can be determined by consideration of possible modified flexagon figures. The single complete cycle is sometimes an only cycle and is sometimes a principal cycle. All the corresponding flexagons are twisted bands, so exist as enantiomorphic (mirror image) pairs. Detailed rules for determination of permissible modified flexagon figures are complicated. Modified flexagon figures can, as described in Section 6.2 for flexagon figures and with due attention to detail, be used to reconstruct compound flexagons. This method was used to design the nets for compound flexagons given below.

9.4   Compound square flexagons

There is one possible flat main position appearance for compound square flexagons and this is the compound ring of eight squares shown in Fig. 9.8. It may be regarded as being formed by the addition of four squares to the main position appearance of a square flexagon (Fig. 3.1). Alternate polygons in the compound ring are attached by opposite sides. There are two distinct types of single complete cycle compound square flexagon. The modified flexagon figures are shown in Fig. 9.9. One has a complete 3-cycle (Fig. 9.9, left) and the other a complete 5-cycle (Fig. 9.9, right).

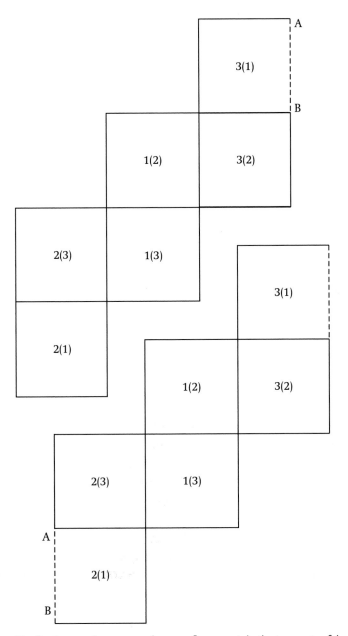

Fig. 9.10 Net for the 3-cycle compound square flexagon. Join the two parts of the net at A-B. Difficult.

The compound ring of eight squares is the first of an infinite family of compound rings made from polygons with an even number of sides in which the number of polygons in the ring is twice the number of sides.

9.4.1 The 3-cycle compound square flexagon

Fig. 9.10 shows the net for the 3-cycle compound square flexagon. This is one of the two possible types of single cycle compound square flexagons.

Fig. 9.11   Tuckerman diagram for the 3-cycle compound square flexagon.

The modified flexagon figure is shown in Fig. 9.9, left. The Tuckerman diagram (Fig. 9.11) shows that there is a cycle of three main positions. These have the appearance shown in Fig. 9.8. There are no subsidiary cycles.

The twist flex, shown by the parallel lines, is used to traverse from one main position to another. To do this the four single leaves at the corners of a main position should be twisted simultaneously. In practice this isn't possible since one doesn't have enough fingers and thumbs, so twists have to be done singly, bearing in mind what one is trying to achieve. For example as assembled the flexagon is in main position 2(1). To flex to main position 3(2) the pats at the sides of the main position have to be unfolded to reveal face 3 and leaves numbered 1 have to be folded together in pairs so as to conceal face 1.

9.4.2   The 5-cycle compound square flexagon

Fig. 9.12 shows the net for the 5-cycle compound square flexagon. This is the second of the two possible types of single principal cycle compound square flexagons. The modified flexagon figure is shown in Fig. 9.9, right. The Tuckerman diagram (Fig. 9.13) shows that there is a cycle of five main positions. These have the appearance shown in Fig. 9.8. There are no subsidiary cycles.

The twist flex is used to traverse from one main position to another. Two types of twist flex are possible. In one type face numbers appear in cyclic order as usual, that is 1(2)-2(3) etc. In the other type both face numbers change, for example 1(2)-3(4). Both possibilities are shown on the Tuckerman diagram (Fig. 9.13). In both types twists have to be done singly, bearing in mind what one is trying to achieve.

9.5   Compound pentagon flexagons

There is one possible flat main position appearance for compound pentagon flexagons and this is the compound ring of 10 pentagons as shown in Fig. 9.14. It is a modification of the regular ring of 10 pentagons (Fig. 9.6) in which the pentagons are separated without changing their orientation, rotated through 180°, and reassembled.

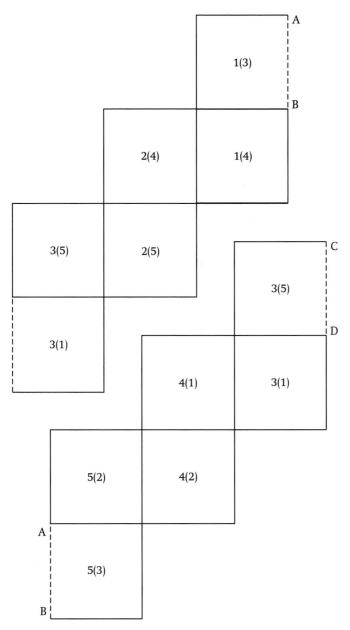

Fig. 9.12 Net for the 5-cycle compound square flexagon. Join the three parts of the net at A-B, C-D. Difficult. (Continued on next page.)

There are four distinct types of single complete cycle compound pentagon flexagon. The modified flexagon figures are shown in Fig. 9.15. One has a complete 3-cycle (Fig. 9.15, top left), one a complete 7-cycle (Fig. 9.15, top right), one a complete 4-cycle (Fig. 9.15, bottom left), and one a complete 6-cycle (Fig 9.15, bottom right).

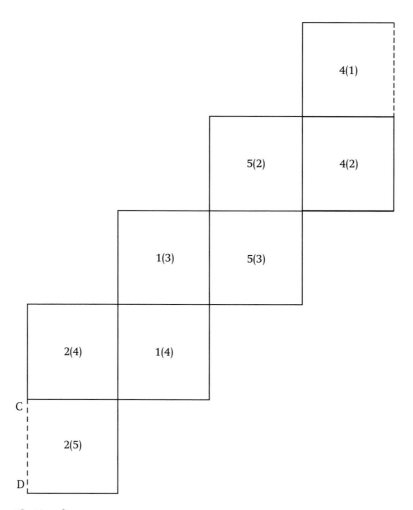

Fig. 9.12   (*Continued*)

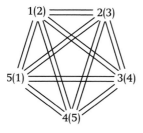

Fig. 9.13   Tuckerman diagram for the 5-cycle compound square flexagon.

The modified flexagon figure shown at top left in Fig. 9.15 shows that there are single leaves at the corners of a flat main position, whereas that shown at bottom left shows that there are single leaves at the sides. This means that a main position link between these two types of cycle

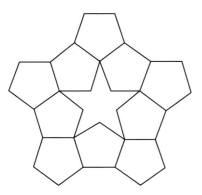

Fig. 9.14 Compound ring of 10 pentagons.

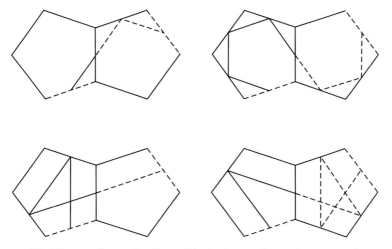

Fig. 9.15 Modified flexagon figures for all possible single complete cycle compound pentagon flexagons.

is possible. Hence with compound flexagons it is possible to have complete cycles with different numbers of main positions and the same main position appearance.

The compound ring of 10 pentagons is the first of an infinite series of compound rings made from polygons with an odd number of sides in which the number of polygons in the ring is twice the number of sides.

9.5.1 The 3-cycle compound pentagon flexagon

Fig. 9.16 shows the net for the 3-cycle compound pentagon flexagon. This is the simplest of the four possible types of single cycle compound pentagon flexagons. The modified flexagon figure is shown in Fig. 9.15, top left. The Tuckerman diagram is the same as that for the 3-cycle compound

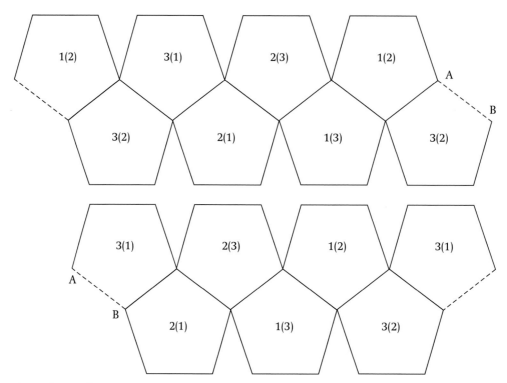

Fig. 9.16   Net for the 3-cycle compound pentagon flexagon. Join the two parts at A-B. Difficult.

square flexagon (Fig. 9.11) and shows that there is a cycle of three main positions. These have the appearance shown in Fig. 9.14. There are no subsidiary cycles.

The twist flex is used to traverse from one main position to another. To do this the five single leaves at the corners of a main position are twisted simultaneously. In practice this is easier than for the 3-cycle compound square flexagon (Subsection 9.4.1) because, if fivefold rotational symmetry is approximately maintained, then little or no bending of the leaves is needed and it is not necessary to hold all the single leaves.

9.6   Compound hexagon flexagons

There are four possible types of flat main position appearance for compound hexagon flexagons. These are a compound ring of 4 hexagons (Fig. 9.17), a compound ring of 6 hexagons (Fig. 9.18) and two types of rings of 12 hexagons (Figs. 9.19 and 9.20). There are two distinct types

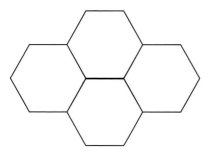

Fig. 9.17   Compound ring of 4 hexagons, type A.

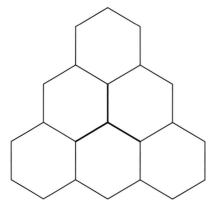

Fig. 9.18   Compound ring of 6 hexagons, type B.

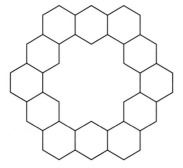

Fig. 9.19   Compound ring of 12 hexagons, type C.

of single principal cycle compound hexagon flexagon for each type of compound ring of hexagons. The modified flexagon figures are shown in Figs. 9.21, 9.22, 9.23 and 9.24. For type A one of them has a complete 5-cycle (Fig. 9.21, left) and the other a complete 7-cycle (Fig. 9.21, right). For type B one has a complete 4-cycle (Fig. 9.22, left) and the other a complete 8-cycle (Fig. 9.22, right). For type C one has a complete 5-cycle (Fig. 9.23, left) and the other a complete 7-cycle (Fig. 9.23, right). Finally, for type D

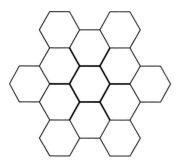

Fig. 9.20   Compound ring of 12 hexagons, type D.

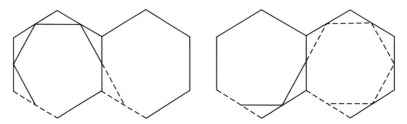

Fig. 9.21   Modified flexagon figures for both possible single principal cycle compound hexagon flexagons, type A.

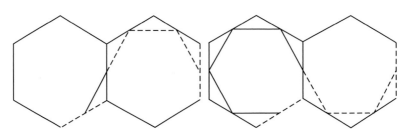

Fig. 9.22   Modified flexagon figures for both possible single principal cycle compound hexagon flexagons, type B.

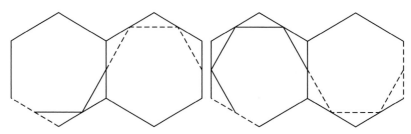

Fig. 9.23   Modified flexagon figures for both possible single principal cycle compound hexagon flexagons, type C.

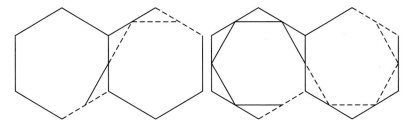

Fig. 9.24 Modified flexagon figures for both possible single principal cycle compound hexagon flexagons, type D.

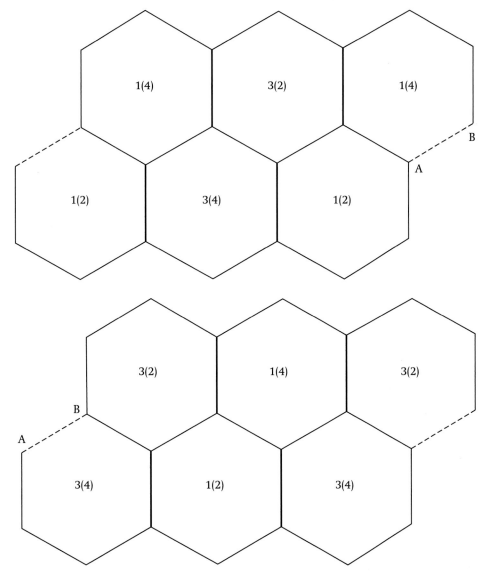

Fig. 9.25 Net for the principal 4-cycle compound hexagon flexagon. Cut along the heavy lines. Join the two parts at A-B. Difficult.

Fig. 9.26   Tuckerman diagram for the principal 4-cycle compound hexagon flexagon.

Fig. 9.27   Intermediate position map of the principal 4-cycle compound hexagon flexagon.

one has a complete 3-cycle (Fig. 9.24, left) and the other a complete 9-cycle (Fig. 9.24, right).

The type C compound ring of 12 hexagons (Fig. 9.19) is the second in an infinite family of compound rings made from polygons with an even number of sides; the compound ring of 8 squares (Section 9.4) is the first. The type D compound ring of 12 hexagons (Fig. 9.20) is a modification of the type C ring in which the hexagons at the corners of the ring have been moved to the inside of the ring without changing their orientation. It is the first of an infinite family of similar modifications of compound rings of polygons with an even number of sides.

9.6.1   The principal 4-cycle compound hexagon flexagon

Fig. 9.25 shows the net for the principal 4-cycle compound hexagon flexagon. This is the simplest of the eight possible types of single principal cycle compound hexagon flexagons. The modified flexagon figure is shown in Fig. 9.22, left. The Tuckerman diagram (Fig. 9.26) for the principal cycle shows that there are four principal main positions. These have the appearance of a type B compound ring of six hexagons as shown in Fig. 9.18. The intermediate position map is shown in Fig. 9.27. In addition to the principal 4-cycle there are two subsidiary 2-cycles. The subsidiary main positions have the appearance of a box like zigzag ring of six hexagons. Each pat in the subsidiary main positions consists of a folded pile of two leaves.

The twist flex is used to traverse from one principal main position to another. To do this the three single leaves at the corners of a principal main position are twisted simultaneously. In practice this is fairly easy since there are only three leaves to be twisted. Patience is needed to find the subsidiary main positions.

10   Distorted polygon flexagons

While writing up earlier chapters it became clear that a fairly systematic treatment of at least some types of flexagon made from distorted polygons was possible. This rather advanced chapter is the result. In order to illustrate some of the enormous range of possibilities nets for several types of distorted polygon flexagon are given, together with descriptions of their dynamic behaviour.

A 'distorted polygon' is a convex polygon derived from a regular convex polygon by changing the shape without changing the number of sides. The leaves of flexagons can be made from any convex polygon, but only a limited range of distorted convex polygons result in flexagons whose paper models are reasonably easy to handle. In practice paper models are most satisfactory when the distorted polygons are semiregular. Distorted polygon flexagons are usually named after the polygons from which they are made.

There are several ways in which the leaves of a flexagon made from regular polygons can be modified to produce distorted polygon flexagons. A distorted polygon can sometimes be regarded as a partially stellated version of a regular convex polygon with a different number of sides. Alternatively, a distorted polygon can sometimes be regarded as a star polygon from which some of the stellations have been removed. If the proportions of the leaves are changed without changing the angular relationships between their sides then, in general, the dynamic behaviour of the flexagon is not affected. Changing the angular relationships between leaf sides does change the dynamic behaviour. Most distorted polygon flexagons are best regarded as variants of either convex polygon flexagons or star flexagons.

10.1   Distorted polygons

A distorted polygon flexagon is a flexagon whose leaves are distorted polygons. A 'distorted polygon' is a convex polygon derived from a regular convex polygon by changing the shape without changing the number of sides. The leaves of flexagons can be made from any convex polygon, but only a limited range of distorted convex polygons result in flexagons whose paper models are reasonably easy to handle. In practice paper models are most satisfactory when the distorted polygons are semiregular. Fig. 10.1 shows some of the leaf shapes which have been used in published nets

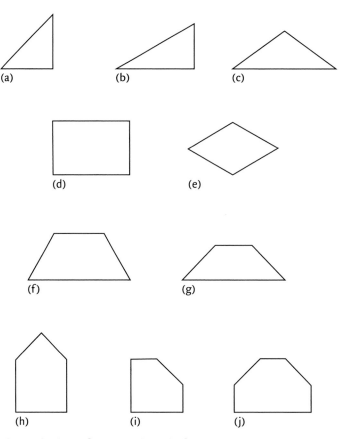

Fig. 10.1   Distorted polygon flexagon polygon leaf shapes: (a) 45°-45°-90° triangle (octagon); (b) 30°-60°-90° triangle (dodecagon); (c) 36°-108°-36° triangle (pentagon); (d) 3:2 rectangle; (e) 60°-120° rhombus (hexagon); (f) 60°-120° trapezium (hexagon); (g) 45°-135° trapezium (octagon); (h) 90°-135°-90°-135°-90° pentagon (octagon); (i) 90°-90°-135°-135°-90° pentagon (octagon); (j) 90°-135°-135°-135°-135°-90° hexagon (octagon).

(Conrad 1960, Conrad and Hartline 1962, McIntosh 2000e, McIntosh 2000f, McIntosh 2000g, McIntosh 2000h, Mitchell 1999).

Distorted polygon flexagons are usually named after the polygons from which they are made. One way of designing a distorted polygon flexagon is to start with a flexagon whose leaves are regular convex polygons and change the shape of the leaves. This usually results in significant changes to the dynamic behaviour. A distorted polygon can sometimes be regarded as a partially stellated version of a regular convex polygon with a different number of sides. For example Fig. 10.2 shows a 60°-120° rhombus as a partially stellated regular convex hexagon. The dynamic behaviour of a flexagon whose leaves are 60°-120° rhombi has some features which are the same as those of a hexagon flexagon. Most of the distorted polygons

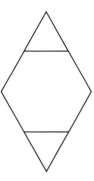

Fig. 10.2   A 60°-120° rhombus as a partially stellated regular convex hexagon.

shown in Fig. 10.1 can be regarded as partially stellated regular convex polygons. Names of polygons which have been partially stellated are shown in brackets in the figure caption. Alternatively, most of them can be regarded as star polygons from which some of the stellations have been removed. Most distorted polygon flexagons are best regarded as variants of either convex polygon flexagons (Chapters 7 and 8) or star flexagons (Section 9.1).

There are several ways in which the leaves of a flexagon made from regular polygons can be modified to produce distorted polygon flexagons. If the proportions of the leaves are changed without changing the angular relationships between their sides then, in general, the dynamic behaviour of the flexagon is not affected. Changing the angular relationships between leaf sides does change the dynamic behaviour. Whatever is done the leaves should superimpose exactly in all the pats of a flexagon. As examples, four possible modifications which result in semiregular quadrilaterals are outlined below.

(a) Change squares to rectangles (Fig. 10.1(d)). The angular relationships between hinges are unchanged.

(b) Truncate equilateral triangles to convert them to 60°-120° trapezia (Fig. 10.1(f)). The angular relationships between hinges are unchanged but an additional side, which could be hinged, is created.

(c) Partially stellate regular convex hexagons to produce 60°-120° rhombi (Figs. 10.1(e) and 10.2). The angular relationships between hinges are unchanged, but the number of sides available for hinging is reduced. The rhombi can be changed to parallelograms without affecting the angular relationships between the hinge lines.

(d) Change the vertex angles of squares to convert them to rhombi. Special cases result if the vertex angles coincide with those of a partially stellated regular convex polygon.

10.2   Some distorted polygon flexagons

Nets for several types of distorted polygon flexagon are given below in order to illustrate some of the enormous range of possibilities. The truncated digon flexagon (Subsection 7.3.1) is a simple example of a distorted polygon flexagon.

10.2.1   A four faced rectangle flexagon

Fig. 10.3 shows the net for a four faced rectangle flexagon. This was derived from the net for the regular single cycle square flexagon (Fig. 6.1) by replacing the squares with rectangles with a length to width ratio of 2. This has no effect on the dynamic behaviour of the principal cycle, which is as shown on the simplified map and the intermediate position map of the irregular single principal cycle square flexagon (Figs. 3.11 and 3.13). These maps also apply to the regular single cycle square flexagon (Subsection 6.1.2).

However, the rectangles used are fairly long and narrow and in two of the box positions a push through flex is possible. Hence, a subsidiary 2-cycle may be traversed as shown in the intermediate position map (Fig. 10.4). This is not possible for the regular single cycle square flexagon. A rectangle is topologically equivalent to a square, but the appearance of the subsidiary 2-cycle means that the intermediate position map is not topologically equivalent to that for the equivalent square flexagon.

10.2.2   A four faced rhombus flexagon with three sectors

Fig. 10.5 shows the net for a four faced rhombus flexagon with three sectors. This has three sectors made from 60°-120° rhombi (Figs. 10.1(e) and 10.2). The sequence of face numbers is the same as on the nets for the regular single cycle square flexagon (Fig. 6.1) and for the regular single cycle truncated square flexagon (Fig. 7.13). There are four main positions which may be traversed in cyclic order but the main positions do not all have the same appearance so this is a pseudo-4-cycle.

A 60°-120° rhombus is a partially stellated regular convex hexagon (Section 10.1, Fig. 10.2) so the four faced rhombus flexagon with three sectors may be regarded as a variant of a hexagon flexagon (Section 8.3). Hexagon flexagons have two sectors. The intermediate position map for the four faced rhombus flexagon with three sectors is shown in Fig. 10.6. Compared with Fig. 8.14, which is the intermediate position map of the regular single principal cycle hexagon flexagon, there are two intermediate positions missing. These are indicated by dots. The four faced rhombus flexagon with three sectors has two subsidiary 2-cycles, two subsidiary

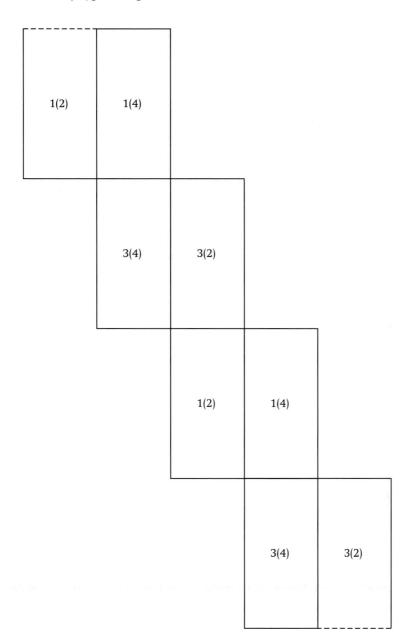

Fig. 10.3 Net for a four faced rectangle flexagon. Difficult.

Fig. 10.4 Intermediate position map of a four faced rectangle flexagon.

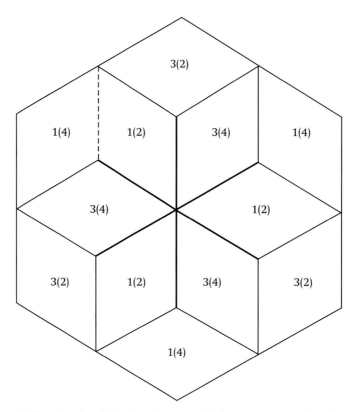

Fig. 10.5 Net for a four faced rhombus flexagon with three sectors. Cut along heavy and dashed lines. Difficult

Fig. 10.6 Intermediate position map of a four faced rhombus flexagon with three sectors.

main positions, 1(2) and 3(4), which belong to the two subsidiary 3-cycles in Fig. 8.14, and two principal main positions, 2(3) and 4(1), which belong to the principal 6-cycle in Fig. 8.14. Subsidiary main positions 1(2) and 3(4) are flat and have the appearance of a six pointed star (Fig. 10.7), but the two principal main positions are skew.

The dynamic behaviour of the four faced rhombus flexagon is similar to that of the regular single principal cycle truncated square flexagon. Their intermediate position maps (Figs. 10.6 and 7.16) are topologically equivalent.

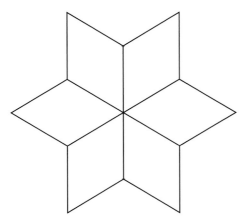

Fig. 10.7   Appearance of subsidiary main positions 1(2) and 3(4) of a four faced rhombus flexagon with three sectors.

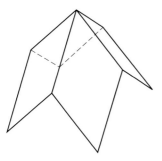

Fig. 10.8   Appearance of subsidiary main positions 1(2) and 3(4) of a four faced rhombus flexagon with two sectors.

10.2.3   A four faced rhombus flexagon with two sectors

Making up the four faced rhombus flexagon described in Subsection 10.2.2 with two sectors instead of three has a significant effect on its dynamic behaviour. To do this use two thirds of the net shown in Fig. 10.5. Subsidiary main positions 1(2) and 3(4) of the four faced rhombus flexagon with two sectors have the appearance shown in Fig. 10.8, and the push through flex is impossible. The push through flex is also impossible for subsidiary main positions 1(3) and 2(4). In consequence the intermediate position map (Fig. 10.9) consists of two disconnected parts. It is not possible to traverse between principal main positions 2(3) and 4(1) without disconnecting a hinge, refolding the flexagon, and reconnecting the hinge. This situation arises for two reasons. Firstly, because the flexagon figure for the four faced rhombus flexagons is a rectangle (Fig. 10.10) and this is not a permissible flexagon figure for a hexagon flexagon (Subsection 8.3.1).

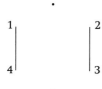

Fig. 10.9   Intermediate position map of a four faced rhombus flexagon with two sectors.

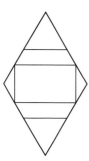

Fig. 10.10   Flexagon figure for the four faced rhombus flexagons.

(In the figure the rectangle is inscribed in a rhombus viewed as a partially stellated hexagon.) Hence it is not possible to traverse between the two principal main positions, 2(3) and 4(1), by using the pinch flex. Secondly the stellations prevent the use of the push through flex so it is not possible to use the push through flex to traverse between intermediate positions 1 and 2, or 3 and 4.

10.2.4   The 3-cycle compound rhombus flexagon

Fig. 10.11 shows the net for the 3-cycle compound rhombus flexagon. This was derived from the net for the 3-cycle compound square flexagon (Fig. 9.10) by replacing the squares with 60°-120° rhombi. The dynamic behaviour is similar to that of the 3-cycle compound square flexagon (Subsection 9.4.1). The same Tuckerman diagram (Fig. 9.11) applies. This shows that there is a cycle of three main positions. These have the appearance of a compound ring of eight rhombi (Fig. 10.12). This has a lower degree of symmetry than the compound ring of eight squares (Fig. 9.8), which makes the use of a modified flexagon figure (Subsection 9.3.1) inappropriate.

The twist flex is used to traverse from one main position to another. To do this the four single leaves at the ends and top and bottom (Fig. 10.12) of a main position should be twisted simultaneously. In practice this isn't possible since one doesn't have enough fingers and thumbs, so twists have to be done singly, bearing in mind what one is trying to achieve.

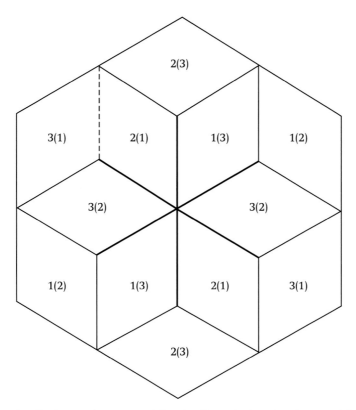

Fig. 10.11 Net for the 3-cycle compound rhombus flexagon. Cut along heavy and dashed lines. Difficult.

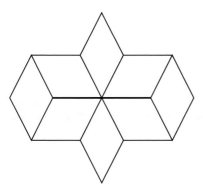

Fig. 10.12 Compound ring of eight rhombi.

For example as assembled the flexagon is in main position 2(1). To flex to main position 3(2) the four pats which consist of a pile of twofolded leaves have to be unfolded to reveal face 3, and the leaves numbered 1 have to be folded together in pairs so as to conceal face 1.

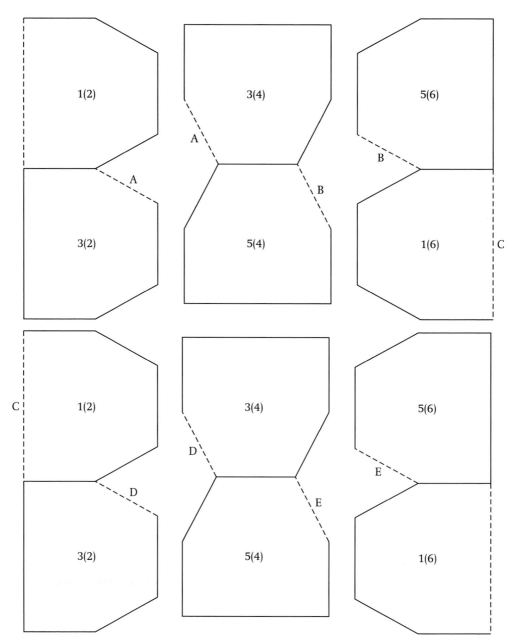

Fig. 10.13 Net for a six faced distorted hexagon flexagon. Join the 6 parts at A, B, C, D, E, keeping the parts the same way up. Fold until face number 1 is visible. Difficult.

10.2.5   A six faced distorted hexagon flexagon

Fig. 10.13 shows the net for a six faced distorted hexagon flexagon. This has two sectors and is made from 90°-150°-120°-120°-150°-90° hexagons. The sequence of face numbers is the same as on the net for the regular

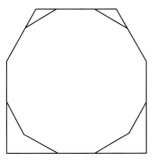

Fig. 10.14   A 90°-150°-120°-120°-150°-90° hexagon as a partially stellated regular convex dodecagon.

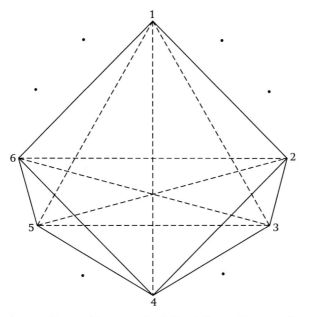

Fig. 10.15   Intermediate position map of a six faced distorted hexagon flexagon.

singular principal cycle hexagon flexagon (Fig. 8.12). A 90°-150°-120°-120°-150°-90° hexagon is a partially stellated regular convex dodecagon (Fig. 10.14) so the six faced distorted hexagon flexagon is a variant of a dodecagon flexagon. Paper models of dodecagon flexagons are impractical, so this variety is not described in Chapter 8.

The side length relationships shown in Fig. 10.14 were modified for the leaves used in the net in order to make the five short sides all the same length and hence make a paper model easier to manipulate. Angular relationships between sides were unchanged so the dynamic behaviour of the flexagon is not affected (Section 10.1).

The intermediate position map is shown in Fig. 10.15. There are six intermediate positions missing, and these are indicated by dots. There are

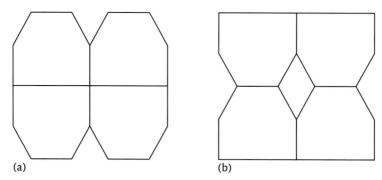

(a)   (b)

Fig. 10.16   Appearance of subsidiary main positions of a six faced distorted hexagon flexagon: (a) 1(2) and 6(1); (b) 2(4) and 4(6).

two subsidiary 2-cycles (subsidiary main positions 1(4) and 4(1), and 2(6) and 6(2)), there is one subsidiary 3-cycle (subsidiary main positions 1(3), 3(5) and 5(1)) and there is one subsidiary 4-cycle (subsidiary main positions 1(2), 2(4), 4(6) and 6(1). There are two subsidiary main positions, 3(4) and 4(5), which belong to a subsidiary 6-cycle, two subsidiary main positions, 2(5) and 6(3), which belong to the subsidiary 12-cycle of the regular single principal cycle dodecagon ⟨12, 12⟩ and two principal main positions, 2(3) and 5(6), which belong to its principal 12-cycle. The intermediate position map is topologically equivalent to that for the regular single principal cycle hexagon flexagon (Fig. 8.14).

The subsidiary main positions in the subsidiary 4-cycle are the only main positions that are flat, and have the two types of appearance shown in Fig. 10.16. Changes in the appearance of main positions in complete cycles are a characteristic of distorted polygon flexagons whose leaves don't have rotational symmetry. Changes in the appearance of main positions while a cycle is being traversed are usually associated with pseudocycles made up of parts from more than one type of cycle, for example Subsection 8.4.2. The apparent paradox disappears if the partial stellations are removed from the leaves. Regular convex dodecagons remain and there is then no change in the appearance of the subsidiary main positions as the subsidiary 4-cycle is traversed. The push through flex is needed to traverse the subsidiary 3-cycle. If the number of sectors were increased from two to three then it would be possible to traverse both the subsidiary 3-cycle and the subsidiary 4-cycle without bending the leaves.

This particular flexagon was designed in an attempt to design a flexagon which included both a complete 3-cycle and a complete 4-cycle. In the event bits of other types of cycle also appeared. The solution presented isn't unique, but it is believed that there are no

Fig. 10.17   A 45°-45°-90° triangle as a regular star octagon with some of the stellations removed.

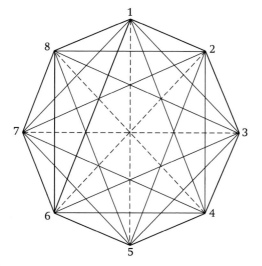

Fig. 10.18   Intermediate position map of a single principal cycle star octagon flexagon.

simpler flexagons which include both a complete 3-cycle and a complete 4-cycle.

10.3   Right angle triangle flexagons

Flexagons made from right angle isosceles (45°-45°-90°) triangles (Fig. 10.1(a)) and with four sectors are sometimes regarded as a variety in their own right (Hilton and Pedersen 1994, McIntosh 2000h, Mitchell 1999). However, a right angle triangle can be regarded as a regular star octagon with some of the stellations removed. (Fig. 10.17). Hence a right angle triangle flexagon is actually a variant of a star octagon flexagon (Section 9.1). The intermediate position map for a single principal cycle star octagon flexagon is shown in Fig. 10.18. There are four subsidiary 2-cycles, two subsidiary 4-cycles, one subsidiary 8-cycle and a principal

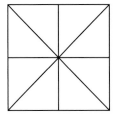

Fig. 10.19 Appearance of principal main positions of right angle triangle flexagons.

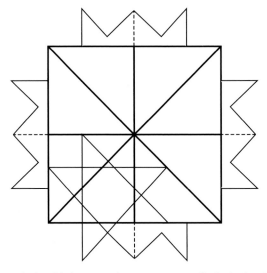

Fig. 10.20 Relationship between the appearances of principal main positions of star octagon flexagons and of those of right angle triangle flexagons.

8-cycle. It isn't possible to make paper models of star octagon flexagons because of interference between stellations. Removal of some stellations makes it possible to make paper models of right angle triangle flexagons. Because of the large number of degrees of freedom fourfold rotational symmetry needs to be maintained during flexing.

The limited number of hinge angles available on the leaves of right angle triangle flexagons means that only pseudo 3-cycles are possible. Two of the main positions in a pseudo 3-cycle are principal main positions belonging to the principal cycle of a single principal cycle star octagon flexagon. These are flat and have the appearance shown in Fig. 10.19. The other is a subsidiary main position belonging to a subsidiary 4-cycle, and is skew. The relationship between the appearances of principal main positions of star octagon flexagons (Fig. 9.1) and of those of right angle triangle flexagons is shown in Fig. 10.20. As with hexaflexagons (Section 4.3) only main position links between cycles are possible. However, there are

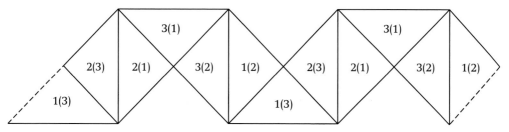

Fig. 10.21 Net for the three faced right angle triangle flexagon. Difficult.

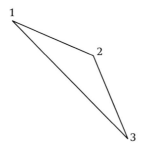

Fig. 10.22 Intermediate position map of a three faced right angle triangle flexagon.

two types of main positions so there are also two types of main position links.

10.3.1 The three faced right angle triangle flexagon

Fig. 10.21 shows the net for the three faced right angle triangle flexagon. The sequence of face numbers is the same as on the net for the trihexaflexagon (Figs. 1.1 and 4.3). The intermediate position map is shown in Fig. 10.22. Compared with Fig. 10.18 there are five intermediate positions missing, and these are indicated by dots. There are two principal main positions, 1(2) and 2(3), which belong to the principal 8-cycle in Fig. 10.18, and there is a subsidiary main position 3(1) which belongs to a subsidiary 4-cycle. Principal main positions 1(2) and 2(3) are flat and have the appearance shown in Fig. 10.19 The subsidiary main position is skew.

As the three faced right angle triangle flexagon is a variant of a star octagon flexagon the flexagon figure (Fig. 10.23) is inscribed in a star octagon. If right angle triangle flexagons are regarded as a variety in their

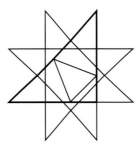

Fig. 10.23 Flexagon figure for the three faced right angle triangle flexagon.

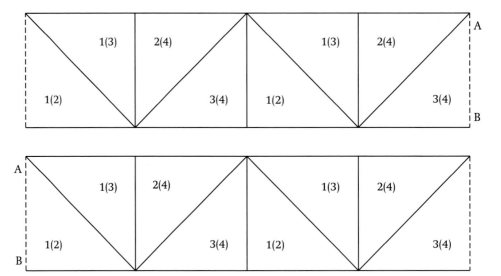

Fig. 10.24 Net for the four faced right angle triangle flexagon with a skew main position link. Join the two parts of the net at A-B. Difficult.

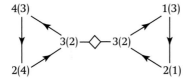

Fig. 10.25 Tuckerman diagram for the four faced right angle triangle flexagon with a skew main position link.

own right then the flexagon figure is inscribed in a right angle triangle. In the figure the circumscribing right angle triangle is picked out in heavy lines. For a flexagon figure to be legitimate the inscribed polygon must connect adjacent sides of the circumscribing polygon (cf. Section 8.2.1). In Fig. 10.23 the inscribed triangle does connects three adjacent sides of the circumscribing star octagon, which confirms that it is a legitimate flexagon figure.

10.3.2 A four faced right angle triangle flexagon

Fig. 10.24 shows the net for a four faced right angle triangle flexagon with a skew main position link. It was formed by linking two three faced right angle triangle flexagons (Subsection 10.3.1) at the subsidiary main positions, which are skew, so it is a skew main position link. The Tuckerman diagram is shown in Fig. 10.25. There are two pseudo 3-cycles. Main positions 1(3), 2(1), 2(4) and 4(3) are flat and have the appearance shown in Fig. 10.19. Main position 3(2) is skew so a snap flex, indicated by the diamond in Fig. 10.25, is needed to traverse from one pseudo 3-cycle to the other. Relaxation of the fourfold rotational symmetry requirement during flexing makes enough degrees of freedom available for this to be done without bending the leaves. It is possible to traverse between main positions 1(3) and 2(1), and between main positions 2(4) and 4(3), by treating the right angle triangle flexagon as if it were a square flexagon and using box flexes.

11 Flexahedra

Four dimensional space is a purely theoretical idea but is nevertheless fascinating. Alicia Boole Stott (1860–1940) was renowned for her ability to visualise four dimensional geometric objects (Coxeter 1963). Most people find this difficult, but it is worth the effort. As an example of the interesting results that can be obtained the dynamic behaviour of Rubik's tesseract, the four dimensional analogue of Rubik's cube, has been investigated (Velleman 1992).

This chapter is a brief introduction to the remarkably rich and largely unexplored topic of 'flexahedra', which are the four dimensional analogues of flexagons. The nets of flexahedra are three dimensional so can be visualised in ordinary space. Sometimes main and intermediate positions of flexahedra are also three dimensional and hence may be visualised. It is possible to generate a flexahedron analogue of any flexagon, and examples are given. The dynamic behaviour of the flexahedron generated is analogous to that of the initial flexagon. There are some flexahedra which are not analogues of flexagons, and one is described. It is of course not possible to make physical models of flexahedra.

Most of the material in this chapter was developed about 30 years ago but has not previously been published. Examples have been chosen to make visualisation as easy as possible.

11.1 Four dimensions

In ordinary three dimensional space it is possible to construct a maximum of three mutually perpendicular straight lines through a given point. In four dimensional geometry it is assumed that it is possible to construct four mutually perpendicular straight lines through a given point. Four dimensional objects are purely theoretical ideas. An introduction to the geometry of four dimensions is given by Coxeter (1963).

It is possible to construct 'flexahedra', which are the four dimensional analogues of flexagons, as ideal mathematical objects. It is of course not possible to make physical models. There do not appear to be any references on flexahedra. Some authors (Conrad and Hartline 1962, Cundy and Rollett 1981, Engel 1969, Hilton and Pedersen 1994, Laithwaite 1980, Schattschneider and Walker 1983) have described three dimensional linkages (Section 3.3) consisting of polyhedra hinged at common edges.

These three dimensional hinged linkages are sometimes referred to as flexahedra.

In ordinary space an ideal plane of infinite extent is two dimensional and divides space into two parts. In four dimensional space an ideal hyperplane of infinite extent is three dimensional and divides four dimensional space into two parts. In ordinary space a square lies in a plane. An observer in ordinary space looking at a square can see its sides and faces, and looking at a cube can see its edges and faces. In four dimensional space a cube lies in a hyperplane. A hypothetical four dimensional observer looking at a cube can see its edges, its faces and two three dimensional hyperfaces.

A flexagon may be regarded as a hinged linkage (Section 3.3). The idea of a linkage can be extended to four dimensions (Macmillan 1950). By analogy with a flexagon (Section 3.1), a flexahedron may be regarded as a band of polyhedra hinged together at common faces. Here, 'hinged' means that the dihedral angle between the two hyperplanes containing the two hinged polyhedra may vary between $0°$ and $360°$ without constraint (Coxeter 1963). A hinge at a face common to two polyhedra is difficult for a three dimensional observer to visualise. The key point is that if one of the hinged polyhedra is fixed in four dimensional space, then the path that may be followed by the other is precisely determined.

11.2   Prism flexahedra

It is possible to generate a 'prism flexahedron' analogue of any flexagon by the following procedure, which is difficult to visualise. Whatever its configuration a flexagon is never more than three dimensional. In four dimensional space a flexagon must therefore lie in a hyperplane. To generate the flexahedron analogue the flexagon is translated through an arbitrary distance perpendicular to the hyperplane. This transforms the leaves of the flexagon into right prisms, hinged at common faces. The dynamic behaviour of the resulting prism flexahedron, including types of flex, is analogous to that of the initial flexagon.

The net for a flexagon can always be laid flat so that, ideally, it lies within a plane. Similarly, the net for a flexahedron can be laid so that it lies within a hyperplane, and in that sense is 'hyperflat'. A hyperplane is equivalent to ordinary space so the hyperflat net of a flexahedron is easily visualised. To generate the net for the prism flexahedron analogue of a flexagon the net for the flexagon is laid flat and then translated through an arbitrary distance perpendicular to its plane. The resulting net for the flexahedron is easily visualised. Similarly, if main or intermediate positions of the initial flexagon are flat then the corresponding main

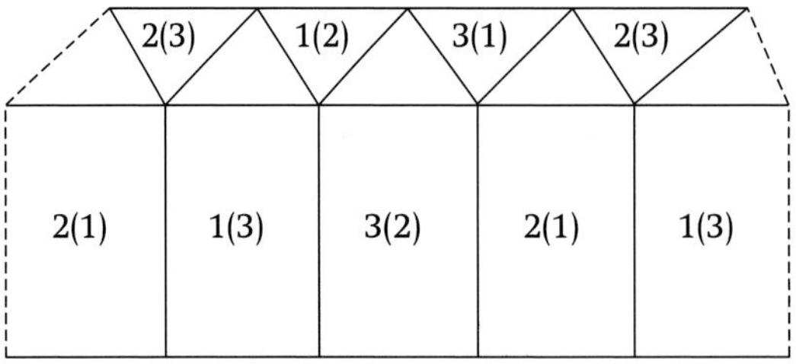

2(3)	1(2)	3(1)	2(3)	
2(1)	1(3)	3(2)	2(1)	1(3)

Fig. 11.1   Net for a three hyperfaced prism flexahedron.

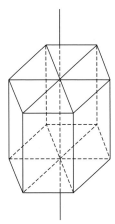

Fig. 11.2   Appearance of a main position of a three hyperfaced prism flexahedron.

or intermediate positions of the flexahedron analogue will also be easily visualised. 'Flexahedron figures', analogous to flexagon figures (Section 3.5), can be constructed for flexahedra.

11.2.1   A three hyperfaced prism flexahedron

Fig. 11.1 shows the net for a three hyperfaced prism flexahedron. This was generated by translating the net for a trihexaflexagon (Fig. 1.1) through an arbitrary distance. In four dimensional space two hyperfaces are visible on the prisms and these hyperfaces may be numbered, as indicated on the net, in the same way as the faces of the leaves of a trihexaflexagon may be numbered (Figure 1.1). Assembly of the net into a prism flexahedron is analogous to the assembly of a trihexaflexagon (Section 1.1). As with the trihexaflexagon (Section 4.2) this prism flexahedron exists in two enantiomorphic (mirror image) forms.

The dynamic behaviour of the three hyperfaced prism flexahedron is analogous to that of the trihexaflexagon (Section 4.2). A main position has the appearance of six triangular prisms about a line, and its outline is a hexagonal prism (Fig. 11.2). An intermediate position is four dimensional so cannot be visualised in ordinary space. The flexahedron figure

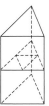

Fig. 11.3   Flexahedron figure for a three hyperfaced prism flexahedron.

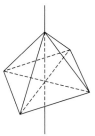

Fig. 11.4   Appearance of a principal main position of a triangular pyramid flexahedron with two sectors.

(Fig. 11.3) is analogous to the flexagon figure (Fig. 3.15) for the initial tri-hexaflexagon. It is an equilateral triangle inscribed in a section through the midpoints of the circumscribing prism faces.

11.3   Pyramid flexahedra

As the name indicates, a pyramid flexahedron consists of pyramids. A main position of a pyramid flexahedron has the appearance of $2n$ right pyramids arranged about a line, with their apices at a common point. The base of each pyramid is a regular convex polygon or a regular star polygon. Two of the slant faces of each pyramid are in common with slant faces of adjoining pyramids. If the height of the pyramids is adjusted so that the dihedral angle between slant faces is $(180/n)°$ then a main position is three dimensional and can be visualised in ordinary space. This is possible provided the vertex angles of the base polygon are less than $(180/n)°$. For other dihedral angles a main position is four dimensional so cannot be visualised in ordinary space.

As an example Fig. 11.4 shows a main position of a triangular pyramid flexahedron with two sectors ($n = 2$). The dihedral angle between slant faces has been adjusted to $90°$ so the main position is three dimensional. The appearance of just the pyramid bases is the same as the appearance of a main position of a triangle flexagon (Fig. 3.4).

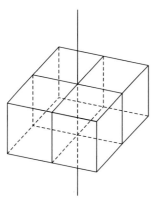

Fig. 11.5   Appearance of a principal main position of a cube flexahedron.

A triangular pyramid flexahedron can be constructed by erecting pyramids on the leaves of a triangle flexagon and hinging together adjacent slant faces. The dynamic behaviour of the triangular pyramid flexahedron depends upon the dihedral angle between slant faces. If this is less than 90° then the dynamic behaviour is analogous to that of the initial triangle flexagon (Section 5.1). In particular to traverse a complete cycle it is necessary to disconnect a hinge, refold the triangular pyramid flexahedron, and reconnect the hinge. If the dihedral angle is greater than 90° then complete cycles may be traversed, but a snap flex is needed to traverse between cycles. When the dihedral angle is 90° then neither hinge disconnections nor snap flexes are needed, and the dynamic behaviour is analogous to that of a hexaflexagon (Chapter 4).

11.4   Cube flexahedra

As the name indicates, a cube flexahedron consists of cubes. A main position has the appearance of four cubes about a line (Fig. 11.5), so can be visualised. In pinch flexing a cube flexahedron a main position is first folded in two about a plane to reach an intermediate position. This has the appearance of two cubes with a common face (Fig. 11.6). An intermediate position may then be unfolded to reach another main position in two different ways. If it is unfolded about the plane parallel to the folding plane then the flex is a 'parallel pinch flex' whereas if it is unfolded about a plane perpendicular to the folding plane the flex is a 'perpendicular pinch flex'.

Cube flexahedron analogues may be generated from flexagons by two different methods. There are some cube flexahedra which do not have flexagon analogues. In the first method a square flexagon is translated

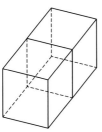

Fig. 11.6   Appearance of an intermediate position of a cube flexahedron.

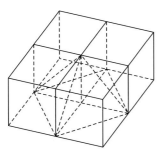

Fig. 11.7   A main position of a triangle pyramid flexahedron transformed into a principal main
position of a cube flexahedron.

through a distance equal to the leaf side length so that the leaves become
cubes. This is a special case of a prism flexagon (Section 11.2). Cycles are
traversed by using the parallel pinch flex. In the second method a triangle
flexagon is first transformed into a triangular pyramid flexahedron with
a dihedral angle of 90° between slant faces (Section 11.3). This triangular pyramid flexahedron is then transformed into a cube flexahedron by
replacing the pyramids with cubes, as shown, for a main position, in
Fig. 11.7. Cycles are traversed by using the perpendicular pinch flex.

It is possible to make flexahedra from any of the other regular polyhedra but main positions are then always four dimensional, and so cannot
be visualised in ordinary space.

11.4.1   A four hyperfaced cube flexahedron

Fig. 11.8 shows the net for a four hyperfaced cube flexahedron. This
was generated by translating the net for the regular single cycle square
flexagon (Fig. 6.1). As with the regular single cycle square flexagon (Subsection 6.1.2) the four hyperfaced cube flexahedron exists in two enantiomorphic forms.

The dynamic behaviour of the four hyperfaced cube flexahedron is
analogous to that of the initial regular single cycle square flexagon

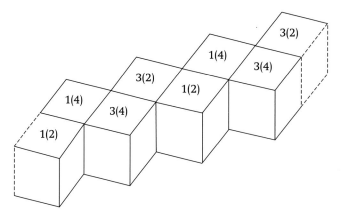

Fig. 11.8   Net for a four hyperfaced cube flexahedron.

Fig. 11.9   Flexahedron figure for a four hyperfaced cube flexahedron.

(Subsection 6.1.2). A main position has the appearance shown in Fig. 11.5, and an intermediate position the appearance shown in Fig. 11.6. Box positions of a cube flexahedron, analogous to box positions of a square flexagon (Subsection 3.4.2), are four dimensional so cannot be visualised in ordinary space. The flexahedron figure (Fig. 11.9) is analogous to the flexagon figure for the regular single cycle square flexagon (Fig. 6.3). It is a square inscribed in a section through midpoints of four circumscribing cube faces. The parallel pinch flex is used.

11.4.2   A four hyperfaced truncated cube flexahedron

If the cubes of the four hyperfaced cube flexahedron (Subsection 11.4.1) are replaced by truncated octahedra then it becomes a four hyperfaced truncated cube flexahedron. A main position of the four hyperfaced cube flexahedron becomes a principal main position of the four hyperfaced truncated cube flexahedron, and has the appearance shown in Fig. 11.10. A cross section through a principal main position has the appearance of a flat ring of four octagons (Fig. 7.14).

The smaller hinges make it possible to turn a box position of the four hyperfaced truncated cube flexahedron inside out by bending the truncated cubes, in a four dimensional sense. This cannot be visualised in

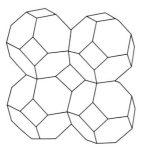

Fig. 11.10 Appearance of a principal main position of a truncated cube flexahedron.

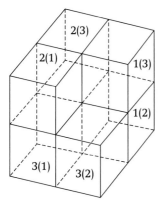

Fig. 11.11 Net for a three hyperfaced cube flexahedron.

ordinary space but Fig. 7.15 provides a three dimensional analogy. Turning the box position inside out makes it possible to traverse two subsidiary 2-cycles as well as the principal 4-cycle. The dynamic behaviour of the four hyperfaced truncated cube flexahedron is analogous to that of the regular single principal cycle truncated square flexagon (Subsection 7.5.1). The intermediate position map (Fig. 7.16) is the same for both. The parallel pinch flex and the box flex are used.

11.4.3 A three hyperfaced cube flexahedron

Fig. 11.11 shows the net for a three hyperfaced cube flexahedron. It was derived from the net for the single cycle triangle flexagon (Fig. 5.1) by using the second method described above. The net is shown as a continuous ring. It has to be disconnected at a hinge and reconnected after assembly of the cube flexahedron. The flexahedron figure (Fig. 11.12) is a triangle inscribed in a triangle connecting three vertices of the circumscribing cube. The perpendicular pinch flex is used.

Fig. 11.12　Flexahedron figure for a three hyperfaced cube flexahedron.

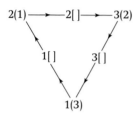

Fig. 11.13　Simplified map of a three hyperfaced cube flexahedron.

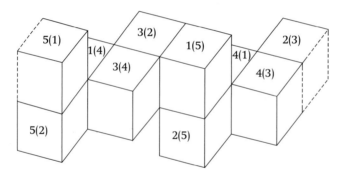

Fig. 11.14　Net for a five hyperfaced cube flexahedron.

Unlike the triangle flexagon (Section 5.1) from which it was derived it is possible to traverse its cycle completely. Hence the dynamic behaviour of the three hyperfaced cube flexahedron is more akin to that of the trihexaflexagon (Section 4.2). In particular the simplified map (Fig. 11.13) is the same for both.

11.4.4　A five hyperfaced cube flexahedron

The cube flexahedra whose nets are shown in Figs. 11.8 and 11.11 may be linked by using a main position link. The method is analogous to that used for square flexagons (Subsection 6.3.1). The net for the resulting five hyperfaced cube flexahedron is shown in Fig. 11.14. The Tuckerman diagram is shown in Fig. 11.15. Cycle A is traversed using the parallel

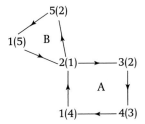

Fig. 11.15 Tuckerman diagram for a five hyperfaced cube flexahedron.

pinch flex and cycle B the perpendicular pinch flex. This flexacube has no flexagon analogue.

The two cycles have different numbers of main positions but main positions in both cycles have the same appearance (Fig. 11.5). This does not usually happen with flexagons (Chapter 8), an exception is certain compound flexagons (Section 9.5).

I always strive, when I can, to spread sweetness and light. There have been several complaints about it.

P. G. Wodehouse, *Service with a Smile*.

References

Chapman, P.B. (1961). Square flexagons. *Mathematical Gazette*, **45**, 192–194.

Conrad, A.S. (1960). *The Theory of the Flexagon*, RIAS Technical Report 60–24. Baltimore, MD: RIAS.

Conrad, A.S. and Hartline, D.K. (1962). *Flexagons*. RIAS Technical Report 62–11. Baltimore, MD: RIAS.

Coxeter, H.S.M. (1963). *Regular Polytopes*, 2nd edn, New York: The Macmillan Company.

Cromwell, P.R. (1997). *Polyhedra*, Cambridge: Cambridge University Press.

Cundy, H.M. and Rollett, A.R. (1981). *Mathematical Models*, 3rd edn, Stradbrooke, Diss, Nfk: Tarquin Publications.

Dunkerley, S. (1910). *Mechanisms*, 3rd edn, London: Longmans, Green and Co.

Dunlap, R.A. (1997/8). Regular polygon rings. *Mathematical Spectrum*, **30**, 13–15.

Engel, D. (1969). Hybrid flexahedrons. *Journal of Recreational Mathematics*, **2**, 35–41.

Feynman, R.P. (1989). *Surely You're Joking, Mr Feynman*, London: Unwin paperbacks.

Gardner, M. (1965). *Mathematical Puzzles and Diversions*, Harmondsworth, Middx: Penguin Books.

Gardner, M. (1966). *More Mathematical Puzzles and Diversions*, Harmondsworth, Middx: Penguin Books.

Gardner, M. (1976). Mathematical games. *Scientific American*, **234**, 120–125.

Gardner, M. (1978). Mathematical games. *Scientific American*, **239**, 18–19, 22, 24–25.

Gardner, M. (1988). *Hexaflexagons and Other Mathematical Diversions*, Chicago: University of Chicago Press.

Griffiths, M. (2001). Two proofs concerning 'Octagon loops'. *Mathematical Gazette*, **85**, 80–84.

Hilton, P. and Pedersen J. (1994). *Build Your Own Polyhedra*, Menlo Park, CA: Addison-Wesley.

Hilton, P., Pedersen, J. and Walser, H. (1997). The faces of the trihexaflexagon. *Mathematics Magazine*, **20**, front cover, 243–251.

Hirsch, R. (1997). *What Is Mathematics Really?*, London: Jonathan Cape.

Hirst, A. (1995). Can you do it with heptagons? *Mathematical Gazette*, **79**, 17–29.

Holden, A. (1991). *Shapes, Space and Symmetry*, New York: Dover Publications, Inc.

Johnson, D. (1974). *Mathmagic with Flexagons*, Hayward, CA: Activity Resource Co.

Kenneway, E. (1987). *Complete Origami*, New York: St Martin's Press.

Laithwaite, E. (1980). *Engineer through the Looking Glass*, London: British Broadcasting Corporation.

Liebeck, P. (1964). The construction of flexagons. *Mathematical Gazette*, **48**, 397–402.

McIntosh, H.V. (2000a). *My Flexagon Experiences*, Puebla, Mexico: Departamento de Aplicación de Microcomputadoras, Instituto de Ciencias, Universidad Autónoma de Puebla.

McIntosh, H.V. (2000b). *REC-F for Flexagons*, Puebla, Mexico: Departamento de Aplicación de Microcomputadoras, Instituto de Ciencias, Universidad Autónoma de Puebla.

McIntosh, H.V. (2000c). *A Flexagon, Flexatube, and Bregdoid Book of Designs*, Puebla, Mexico: Departamento de Aplicación de Microcomputadoras, Instituto de Ciencias, Universidad Autónoma de Puebla.

McIntosh, H.V. (2000d), *Heptagonal Flexagons*, Puebla, Mexico: Departamento de Aplicación de Microcomputadoras, Instituto de Ciencias, Universidad Autónoma de Puebla.

McIntosh, H.V. (2000e). *Hexagon Flexagons*, Puebla, Mexico: Departamento de Aplicación de Microcomputadoras, Instituto de Ciencias, Universidad Autónoma de Puebla.

McIntosh, H.V. (2000f). *Pentagonal Flexagons*, Puebla, Mexico: Departamento de Aplicación de Microcomputadoras, Instituto de Ciencias, Universidad Autónoma de Puebla.

McIntosh, H.V. (2000g). *Tetragonal Flexagons*, Puebla, Mexico: Departamento de Aplicación de Microcomputadoras, Instituto de Ciencias, Universidad Autónoma de Puebla.

McIntosh, H.V. (2000h). *Trigonal Flexagons*, Puebla, Mexico: Departamento de Aplicación de Microcomputadoras, Instituto de Ciencias, Universidad Autónoma de Puebla.

McLean, T.B. (1979). V-flexing the hexahexaflexagon *American Mathematical Monthly*, **86**, 457–466.

Macmillan, R.H. (1950). The freedom of linkages. *Mathematical Gazette*, **34**, 26–37.

Madachy, J.S. (1968). *Mathematics on Vacation*, London: Thomas Nelson and Sons Ltd.

Maunsell, F.G. (1954) The flexagon and the hexaflexagram. *Mathematical Gazette*, **38**, 213–214.

Mitchell, D. (1999). *The Magic of Flexagons Paper: Manipulative Paper Puzzles to Cut Out and Make*, Stradbrooke, Diss, Nfk: Tarquin Publications.

Neale, R. E. (1999). Self-designing tetraflexagons. In *The Mathematician and Pied Puzzler*, Ed. E. Berlekamp and T. Rodgers, pp. 117–126. Natick, MA: A. K. Peters Ltd.

Oakley, C.O. and Wisner, R.J. (1957). Flexagons. *American Mathematical Monthly*, **64**, 143–154.

O'Reilly, T. (1976). Classifying and counting hexaflexagrams. *Journal of Recreational Mathematics*, **8**(3), 182–187.

Pedersen, J.J. and Pedersen, K.A. (1973). *Geometric Playthings to Color, Cut and Fold*, San Francisco: Troubadour Press.

Sawyer, W.W. (1943). *Prelude to Mathematics*, London: Penguin Books.

Schattschneider, D. and Walker, W. (1983). *M. C. Escher Kaleidocycles*. Stradbrooke, Diss, Nfk: Tarquin Publications.

Sloane, N.J.A. and Plouffe, S. (1995). *The Encyclopedia of Integer Sequences*, San Diego: Academic Press.

Taylor, P. (1997). *The Complete? Polygon*, Ipswich: Nattygrafix.

Velleman, D. (1992). Rubik's tesseract. *Mathematics Magazine*, **65**, 27–36.

Wenninger, M.J. (1971). *Polyhedron Models*, Cambridge: Cambridge University Press.

Wheeler, R.F. (1958). The flexagon family. *Mathematical Gazette*, **42**, 1–6.

Flexagon index

Subject index